A Barrister for the Earth

MONICA FERIA-TINTA

A BARRISTER FOR THE EARTH

TEN CASES OF HOPE FOR OUR FUTURE

faber

First published in 2025
by Faber & Faber Limited
The Bindery, 51 Hatton Garden
London EC1N 8HN

Typeset by Typo•glyphix, Burton-on-Trent, DE14 3HE
Printed and bound by CPI Group (UK) Ltd, Croydon, CR0 4YY

A CIP record for this book
is available from the British Library

ISBN 978–0–571–38636–9

Printed and bound in the UK on FSC® certified paper in line with our continuing
commitment to ethical business practices, sustainability and the environment.
For further information see faber.co.uk/environmental-policy

Our authorised representative in the EU for product safety is
Easy Access System Europe, Mustamäe tee 50, 10621 Tallinn, Estonia
gpsr.requests@easproject.com

2 4 6 8 10 9 7 5 3 1

To Klemens Felder, in gratitude,
for his love for, and luminous teachings on,
the natural world.

Contents

Introduction ix

1. The Wayúu and the Largest Open-Pit Coal Mine in Latin America 1
2. Myanmar: Southeast Asia's Last Free-flowing River 37
3. The Dark Business of Light in the Land of Birds 65
4. All Mankind is One: The Camisea Gas Project in Peru and Non-contacted Tribal Peoples 101
5. Sinking Islands: The Torres Strait Islanders 125
6. West Timor: The Montara Oil Spill Case 153
7. The Heart of the Earth: La Línea Negra 181
8. The Masewal and the Constitutionality of Federal Mining Law in Mexico 201
9. The Rights of Nature Case: Los Cedros Cloud Forest 217
10. Into the Deep Blue 247

Epilogue 269
Notes 279
Acknowledgements 333

Introduction

'You cannot predict the seasons any longer,' observed one of my clients, an indigenous Melanesian inhabitant of the Torres Strait Islands. His reality had become dystopic. Life on Boigu Island, the most northerly inhabited island in the Torres Strait, close to Papua New Guinea, had been determined by the seasons since ancestral times. That certainty is now gone. To make matters worse, every year a piece of the island is being 'eaten' by the sea. The Torres Strait Islands are sinking, and it's because of climate change.

It's not just in the Torres Strait. On the opposite side of the world, on the northernmost peninsula of South America, in a land of water scarcity, waterways are diverted to extract coal from riverbeds, all to power Western civilisation. This is not some kind of 'magic realism' – it is happening as I write. In the home of the Wayúu indigenous peoples, in remote Guajira, Colombia, the main waterways have been polluted by open-pit coal extraction and the diversion of rivers is altering the climate. Once reliable, rain patterns have now shifted. But the mine expansion continues regardless and at an unremitting pace. Once a hidden gem of the Northern Caribbean coast, La Guajira is now characterised by cracked parched earth.

But it's not just the Torres Strait and Colombia. Natural gas is shipped to the UK from the Peruvian Amazon rainforest. This gas project has relentlessly penetrated protected rainforest, home to Amazonian tribes in voluntary isolation. As a result, today those tribes are suffering from mercury poisoning.

'In fifteen kilometres of river, they want to build three dams,' I was told by a Mayan Ch'orti', who was defending the Jupilingo River, near the border with Honduras, from a hydroelectrical project that threatened their ancestral land. And this sort of threat is not unique to Guatemala. A year earlier, I had witnessed Asia's last free-flowing river, the Salween in Myanmar, face the same predicament.

In yet another part of the world, in the Timor Sea, one of the world's worst offshore oil rig accidents has taken place. The spill happened in Australian waters, but the harm it has caused respects no boundaries. At least 23 million litres of crude oil has poured into the ocean, destroying the lives and livelihoods of West-Timorese fishermen and seaweed farmers.

Elsewhere, in Colombia, indigenous peoples, such as the Kogi, Arhuaco, Kankuamo and Wiwa, struggle to protect an area in Sierra Nevada de Santa Marta they call the 'Heart of the Earth'. This area has been identified as being vital for the ecological balance of the world: it is richly biodiverse, with over 700 bird species and 3,000 plant species.

At times there are no human beings directly affected. In Ecuador, in the Los Cedros cloud forest near the equator line, endemic species that cannot be found anywhere else in the world are threatened with extinction by a gold-mine exploration.

These are all snapshots from my practice as a barrister in London.

I found myself acting, first, in cases of indigenous peoples facing uphill, Sisyphus-like battles, which went on to set ground-breaking precedents in law – until one day, I realised that my client was no longer a human being. I found myself pleading directly for the rights of numerous species to simply exist. I had become a barrister for the Earth.

Can a planet have legal rights? Could they be defended in a court of law? And who or what are we referring to when we refer to the right to *life*?

In recent years, a quiet revolution has been taking place. Ordinary people have turned to courts around the world seeking justice for environmental damage: melting glaciers, sinking islands, extinction of species, and more. This is not a marginal development, and the legal cases are not ordinary ones. They prompt a reinterpretation of the law itself, one that is likely to impact the life of every single person around the world. These cases will secure our future. They are landmarks signalling that we are at an important juncture: the beginning of a shift in consciousness, and the dawning of a new era.

I know this well. I started to work on environmental cases when I began my practice as a barrister in 2014, though my experience as a public international law expert goes back to 1997. I have been at the crest of this wave of cases since 2016, when I was invited to consider how, in the absence of an inbuilt legal mechanism, the newly adopted Paris Agreement could be enforced. I attempted to provide a pathway using international law, as it was clear to me that the law was a powerful tool in the battle to redress environmental harm and to fight ecological degradation and climate change.

From that beginning, developing legal arguments in theoretical form, I took these arguments to the courts. In 2018 I took up the case of the Torres Strait Islanders against Australia, litigating from my flat in London all the way to the United Nations Human Rights Committee, in Geneva. In September 2022, the Committee delivered a ground-breaking decision: the first international dictum holding a state responsible for lack of action in addressing climate degradation. This was a momentous decision. It was a concrete demonstration that the law could have a definitive role in protecting the Earth. In 2018 I also pioneered legal arguments to bring a climate case in the International Tribunal for the Law of the Sea, at a time when people thought this was unarguable or not possible. Most thought that climate change was not a matter for courts.

Little known and with its headquarters in a white building in Hamburg, this tribunal heard exactly such a case in September 2023, delivering a ground-breaking Advisory Opinion on climate and the protection of the ocean and marine species. When the decision came out, I was ecstatic. All the effort and long hours developing pleadings finally made sense. It was precisely what I had envisaged and hoped for.

In this book, I share that vision about how the law can play a definitive role in protecting the Earth much more widely, and in doing so I would like to demonstrate that we already have the tools to solve the apparently unsolvable crises of our times.

In his book *Irreplaceable*, Julian Hoffman warned readers about those places where the loss of nature is 'real, visceral and imminent'.[1] In recent years I have discovered such places for myself, as a barrister arguing in front of constitutional courts and international organs on the meaning of the word 'imminent', and even the meaning of the word 'life'. These are not topics that have always preoccupied me, however. I grew up in Lima, a grey city that failed to impress Darwin when he visited while on board the *Beagle*: he described it as a place with a 'thick drizzling mist, which was sufficient to make the street muddy and one's clothes damp'.[2] My parents had migrated there from the countryside at a young age. I was therefore born in Lima, and lived there happily in a concrete jungle.

I cannot pinpoint precisely when my own consciousness of the natural world properly came into being. It evolved slowly. I survived atrocities in my country of origin and arrived in England as a refugee. There I found healing, sitting in my first garden, observing blue tits coming to bathe in the gutters of an old house. Watching this magical act had an effect no psychologist could have achieved. It reconciled me with the world. In 2001, I saw a documentary called *Winged*

Migration by Jacques Perrin, Jacques Cluzaud and Michel Debats. This film gave me a new insight into my own migration. Compared to the birds, mine had been a relatively uncomplicated experience. From that point on I have looked at birds with great respect. They taught me about resilience and how to be brave.

My understanding of the law has always been informed by what reality has taught me. On a trip to Mexico in the early 2000s, I visited a village near Michoacán that no longer had running water. I observed that the only water available was in bottles. The indigenous people told me that, in the past, water from natural springs directly supplied their homes. Then a company had come, bottled the spring water and sold it in plastic bottles to those very same people. The owner of that company was Coca-Cola. This was the first time that I began to realise what was going on with water access around the world, but it would not be the last.

In *Gravity and Grace*, the philosopher and mystic Simone Weil observed that 'in the intellectual order, the virtue of humility is nothing more nor less than the power of attention'.[3] It is our power of attention, our shift in consciousness, that will ensure the survival of the natural world – and our own. My attention began to sharpen up when, in 2015, at the Middle Temple, I watched the Shakespearean actor Mark Rylance's re-enactment of the sixteenth-century debate of Bartolomé de las Casas and Juan Ginés de Sepúlveda on whether indigenous peoples have souls. The re-enactment triggered a reflection within me on the relevance of that debate today, when indigenous peoples around the world are fighting for survival and the protection of their ancestral land, some of the last areas of wilderness that exist in this world.

I have since become involved in many cases related to indigenous peoples, including that of the uncontacted tribes in the

Nahua-Nanti Reserve, who are fighting against the Camisea gas project in the Amazon. This project has relentlessly penetrated protected rainforest, home to Amazonian tribes in voluntary isolation. The tribes became contaminated with mercury poisoning. There is often an assumption that such realities can be self-contained in some far-away lands. During my practice as a barrister, however, I have come to understand that this is not usually the case. When I learned that natural gas from the Nahua-Nanti Reserve was shipped to the UK in a tanker owned by Shell, it brought home to me, literally, that what was happening thousands of miles away could painfully, albeit indirectly, implicate us all.

This book lays bare the peril faced by the natural world and tells the story of those fighting against its large-scale destruction. It describes ten cases in which I have acted, and each of them tells a story. They are compelling stories in their own right, and each exposes a dimension of the issues humanity is facing today in its relationship with the Earth, and the remedies the law is able to provide.

This is a book I'm writing as a matter of urgency. You need to know what is happening in remote, biodiverse, ecologically rich parts of the planet, because it affects you, and it possibly also implicates you in some way. When a rainforest dies, or a species goes extinct, or an ocean is despoiled, it is inextricably connected with you, with all of us. The indigenous peoples I have represented taught me about this truth, but this concept is not unknown to the West. The naturalist Alexander von Humboldt realised, after visiting El Chimborazo, in the Andean foothills, that 'Nature is a living whole [. . .] a web in which everything is connected'. 'One single life,' he realised, 'had been poured over stones, plants, animals, and humankind', and 'even the atmosphere carried the kernels of future life – pollen, insect eggs and seeds.'[4] I have come to understand this principle, through my work. When I fought the legal battle for the Torres Strait Islanders, I started by studying the situation in the Arctic, with the failed but important legal battle of the Inuit years earlier, who had warned us that what was happening in the Arctic (the melting of their world) would not stay in the Arctic. The effects, I was to learn, reached all the way to people in the Pacific.

When the kernels of life are at threat because of human intervention, it is humankind's responsibility. And, for those of us who live in the West, it very often happens in our name, and for services we use. I myself once assumed that the destruction happening somewhere else did not implicate me. But it did. While the chapters in the book follow mostly a chronological order, it opens out of sequence with the case relating to an open-pit coal mine. I've opened with it as this was the moment in which I properly comprehended, in its fullest dimension, the very premise of our current 'civilisation'. It shocked and deeply disturbed me. This case was pivotal; and it marked the moment in which I decided to write this book. I wanted to raise attention, to tell these stories as I saw them to prompt your understanding, your awareness, and to try to protect all our futures. I needed you to understand that the coal mine in La Guajira is connected to the rising of the seas in the Torres Strait. I wanted you to understand that the plight of the people in voluntary isolation in the Peruvian Amazon is fundamentally connected to us and our idea of 'development'.

But this book is not simply about the destruction of our natural world. It follows some of the emblematic cases I've worked on and represented as a barrister with a particular purpose. It tells the story of how we can defend nature, how we can fight back on its behalf and how we can win. It follows a very personal path, I confess, as I have evolved myself along the way as an advocate. For example, in how I managed to pioneer a new way of thinking about the law and interpreting pre-existing legal frameworks to ensure that the climate found its way into international court rooms. In how I followed my own consciousness, dared to stand and speak my mind, and win in matters that, while appearing complex, just needed the laser focus of a legal mind to argue the case for the Earth. But while I hope to prompt further successful climate and environmental litigation, that is not my sole goal. What I want to see, and what I believe is already beginning, is a reimagining of the way we exist on this planet, a fundamental reimagining of the way we think.

That's why I'm writing this book. This is a window into reconceiving a way to exist on this planet. The cases I have worked on forced me to pay attention to the natural world as something much more than 'a resource'. I came to understand the way those closely connected with nature see the sacredness of all that is living, including rivers, mountains and forests. The Los Cedros cloud forest, near the equator line, was a 'client', but it was also a teacher.

It was in June 2020 that I found myself acting in a case relating to this cloud forest in Ecuador, in which live species that cannot be found anywhere else in the world, and which were being threatened with extinction by a gold-mine exploration. When working on the claim, during lockdown, in my kitchen in London, I became immersed in the identity of this forest ecosystem. Simone Weil would have said 'to be in direct contact with nature [. . .] is the only discipline'.[5] As I found myself engrossed in the living world of the cloud forest, I realised I was no longer advocating for people, but advocating directly for Nature, for the diversity of life that created and sustains us. It was then that I realised I was effectively acting as a 'barrister for the Earth'. I appeared before the Constitutional Court of Ecuador arguing for the right of the spectacled bear, a variety of monkeys, fungi, frogs, and the entire cloud forest itself, to exist. Hundreds of scientists appeared before the court as well, giving evidence on the biodiversity value of this place.

I realised that this was a turning point. Something similar to what our elders may have experienced during the Enlightenment. The Court delivered a full reasoned judgment – the first of its kind in the world – on the Rights of Nature. To my mind one of the most important dictums of the century. This story, the climax of this book, is captured in Chapter Nine.

The cases I have worked on have led me to believe not only in the transformative power of the law, but also that a shift in paradigm is needed.

In the 1948 Universal Declaration of Human Rights, the right to life as a legal notion became anthropocentric, human centred. This landmark moment somehow also marked the loss of indigenous visions of the world, where all that is living has equal rights.

Did we get it wrong? Has a system of law that only respects the right to life of human beings missed the point regarding the true protection of life? Isn't the destruction of ecosystems contrary to the right to life? Shouldn't the law protect life in *all* its forms?

In this book, I question our current notion of civilisation, even of 'progress', if it is premised on the death and extinction of entire species. The shift is fundamentally one of reason. One can see it taking place in new constitutional processes and emerging in the first international dictums dealing with the environment in the twenty-first century, paving the way to a more evolved system of law and of living our lives on Earth.

ONE

The Wayúu and the Largest Open-Pit Coal Mine in Latin America

The Wayúu Women: The Ouutsü or Dreamers

How do you take on three multinational energy giants from your living room? How do you maintain the laser-like focus of a marks-woman for weeks at a time?

That is precisely what I did. Was it indignation? Was it anger? Was it the loss of innocence?

It is simple. As a barrister, you are instructed. You do your best for your client. You don't have to agree with him or her, but you are not merely 'a hired brain'. You ultimately serve the rule of law. That is my motivation. Yet this case shook me to the core. I never saw the world in the same way again.

I do not know how they found me, the Ouutsü women. But they did. In Wayúu culture, only women (not men) can 'dream'; that is, provide spiritual guidance and healing through the interpretation of messages, which come to them in dreams. Did the idea of a lawsuit beyond Colombian borders come to them in dreams? Did they visualise an international lawyer who could argue their case for them from far-away lands in a dream? The world of dreams in Wayúu culture is just as important as conscious reality, and while the Wayúu women are dreaming, the deity of sleep, Lapü, visits them to transmit important messages. But the Wayúu women's dreams have been disturbed in recent times.

Reaching out to me to denounce their current conditions was a last resort, a desperate bid to try to ensure their survival. They confront an injustice that no Palabrero or Pütchipü'üi (a wiseman that resolves conflicts by peaceful means in the Wayúu world) has been able to mediate. For four decades the Wayúu's plight had been invisible to the West. But that was about to change.

The Wayúu are indigenous peoples scattered across more than 15,000 km^2 of La Guajira in the northernmost part of Colombia and northwest Venezuela.[1] The Guajira Peninsula is the driest and hottest place in Colombia. Reportedly, they settled there some 10,000 years ago and survived 'by specialising their hunting and fishing skills, and using rare fresh-water sources for horticulture', and they became legendary for their resistance to the conquistadors; the Spanish, who colonised much of Latin America, failed to conquer the Wayúu or steal their territory.[2] To this day, the Wayúu are the most numerous of the indigenous peoples in Colombia. Their language is Wayúunaiki.

The Wayúu have a matriarchal culture.[3] How fascinating is that? This is not only because the majority of Wayúu women are the main economic providers, but also because women are the carriers of ancestral knowledge. The Wayúu social order is based on a desirable state of harmony between society, individuals and the natural environment,[4] and the Ouutsü – the Wayúu women who possess the gift of precognitive 'dreams' – also possess special attributes with which to connect to the natural world, including ancestral knowledge regarding the use of plants for medicinal purposes. They are the spiritual authorities.

Despite resisting the conquistadors, for the last four decades this vast landscape – the ancestral land of the Wayúu – has been occupied by Carbones del Cerrejón Ltd.[5] It is one of the largest open-pit coal mines in the world, the largest in Latin America, covering an area of approximately 69,000 hectares. It was hard for me to assimilate, at first, its sheer size. When I took up this case, it was owned by three multinational companies: Glencore (a Swiss Company), Anglo American (a British Company) and BHP Billiton (an Australian Company).

What I was about to learn was deeply disturbing. It is a story of brutal dispossession and of massive contamination at a scale I thought unimaginable. It is a story that we don't often see occupying the headlines in newspapers in the West. In fact, I knew nothing about La Guajira until a message reached my inbox.

An urgent message

During the pandemic I was contacted, as a matter of urgency, by Rosa María Mateus, a lawyer working for one of the most respected human rights organisations in Colombia, the José Alvear Restrepo Lawyers' Collective (CAJAR). She was seeking to instruct me on behalf of members of the Wayúu community of El Provincial. The Wayúu of El Provincial, she told me, lived next to a pit of the Carbones del Cerrejón mine and they had sought relief from the ongoing contamination all the way to the Constitutional Court. A recent Colombian Constitutional Court ruling had described the situation of the Wayúu as 'characterised by indigenous communities that present a high degree of vulnerability and large-scale mining exploitation that puts their environment and health at risk'.[6]

Over the phone, Rosa María briefly outlined the case. The miracle of long-distance communication; we were speaking with each other despite lockdowns and time-zone differences. 'The situation has become more urgent than ever; the ongoing humanitarian crisis existing in the Guajira is now exacerbated by the Covid-19 health emergency,' she told me. The dust of the mine (which operates twenty-four hours a day) made the Wayúu more vulnerable to Covid. 'There is no running water in La Guajira, Mónica,' she emphasised. 'They receive potable water in bottles. The rivers and streams are otherwise contaminated.' This lack of water aggravated their risk of Covid infection. Rosa María's speech was like a stream of consciousness. 'There is no access to hospitals, or medical attention,' she continued.

Even in developed countries, like the one in which I was living, people were fighting to breathe and hospitals were struggling to cope. During the first lockdown, in La Guajira, catching Covid was particularly dangerous. In all likelihood you would simply have no access to oxygen in a hospital, if it was required, and your vulnerability was heightened because your respiratory system was already compromised. The Wayúu were vulnerable. At the time Rosa María contacted me, a traditional Wayúu leader had just passed away because of Covid. The Wayúu were worried about his second wake. Rosa María explained: 'The passing of a person is a big deal for the Wayúu.' After the first burial, they have a second burial. In the first burial the Wayúu bury the deceased following their own rituals. The deceased is wrapped in her/his hammock, interred along with material items necessary for daily life. The Wayúu gather and cook in large quantities for those who are mourning. Years later, an elderly member of the family, a woman, is in charge of removing the remains of the corpse and cleaning them to later place them in a clay pot. That is the real farewell. But they can't practice any of these ceremonies if people die of Covid. To avoid contagion, people could not be near those infected with Covid; they died alone.

Rosa María told me that all possible efforts to address the pollution produced by the mine had been exhausted all the way to the highest court in Colombia, the Constitutional Court. A peculiarity of this case was, indeed, that it had been litigated broadly. The profusion of court judgments was staggering. The harms caused by the mine operation had been fully documented and acknowledged by Colombian courts on multiple occasions, but, despite this, their judgments went unheeded by the mine.

I told Rosa María and Luz Marcela Pérez, a young lawyer from CAJAR assisting in the case, to forward all the background information and evidence. I was in.

It was clear to me that their case needed to be brought to more worldwide attention. I needed to bring their situation to the attention of an international organ, in order to denounce the human rights and environmental violations being committed and get urgent measures granted. I advised on the possible route to follow: namely, to seek an urgent intervention from the United Nations, and my instructions were formalised. The aim was to complete an application on behalf of the Wayúu within two weeks.

Unless you are a lawyer, it may be difficult to understand the moment when you confront a case such as this one for the first time. You are presented with a crisis, something you know nothing about, except that it is harming people right then and there, and – confronted with masses and masses of papers – you have to digest, unravel and identify what matters, and find the key to resolving it.

At the time, I was in lockdown in London. Middle Temple Young Barristers' Association had organised a collective viewing of a documentary online as a way to perhaps help mitigate the isolation of members. I signed up. This screening was the only interruption to the seclusion I had during the first lockdown in London. The film was *Dark Waters*, which follows a lawyer's legal fight taking on the chemical giant DuPont after discovering that the company was polluting drinking water with harmful PFAS, or 'for ever chemicals'. There was a scene in the film that stuck in my brain, because I felt exactly the same at the start of the case I had just taken on myself. The lawyer was alone, in an artificially lit room that was full of boxes containing papers relating to the case – he had to read through the evidence box by box. It was 'a job for ants', as I call it:

an overwhelming amount of data that requires humble, patient, dedicated work. I felt exactly the same as that lawyer. I was there, with masses and masses of judgments and reports, grappling with the reality of La Guajira.

Most people I knew were scarcely coping with isolation. Instead, I was at my best, making the most of the quiet time at my disposal, free from distractions. My sole focus for two weeks was this case.

I threw myself into the evidence and digested copious amounts of information. Through the endless pages I felt the heat of La Guajira, the lack of water, the carbon dust, the frustration. I obsessively listened to *Los Caminos de la Vida* (The Paths of Life) by Los Diablitos, a Vallenato (an upbeat type of music from the Atlantic coast of Colombia), whenever I needed to stop. I worked almost feverishly, with uninterrupted concentration, sometimes for over fourteen hours a day. At times the facts astounded me. As I went deeper, uncomfortable truths were brought to light; they revealed a connection between the energy we enjoyed in the West and the disturbing reality the Wayúu faced – a revelation that made me ask questions I had never asked myself before. Where does the energy that powers my computer really come from? What is the money I deposit in my account at the bank used for?

Cerrejón, once a sacred mountain

At the outset, I need to stress the following. I have called the mine by its official name, Carbones del Cerrejón Ltd, and not Cerrejón, as it is often called. This is a deliberate choice, in order not to perpetuate a cultural misappropriation. The commonly used name was taken from that of a sacred mountain – Cerrejón – in the ancestral land of the Wayúu people. This misuse, I learned, is deeply resented by the Wayúu. To them it is a particularly perverse cultural misappropriation, one that, sadly, is far from unusual. For example, in Mexico City, the Spaniards, the colonisers, built their cathedral on top of the former Aztec sacred precinct. It is a symbol of obliteration. I erase you. I obliterate your gods. I impose my power.

In the past, the Cerrejón mountain was used to determine the weather forecast; the clouds would rest on Cerrejón, for example, when it was going to rain. And above the mountain was a mystery: a sacred lagoon guarded by white monkeys, howler monkeys and black monkeys. The mountain has a spiritual, cultural, social and environmental value, which little by little has been taken from the Wayúu people by the activities of the mine. The guardian monkeys are, little by little, disappearing.

In the Wayúu's words, 'with the arrival of mining, they stripped us of our beliefs, since the sacred sites were taken away; they wiped out vegetation, animals, and waters. There was also an intrusion into our culture and fragmented our communities, families and friendships forged over 400 years ago.'[7]

The territory of La Guajira is also home to afro-Colombian communities. The Wayúu and afro-Colombian communities have, for years, been struggling against forced relocation, health issues, environmental degradation and the destruction of their waterways, all of which can be traced back to the mine. According to rulings of an administrative court in Riohacha, and other courts, Carbones del Cerrejón's mining operations directly impact a population of more than 300,000 people, in an area of 200 km^2.[8] As a result of the mine's activities, thirty-five communities have been displaced from their lands and seventeen waterways have dried up.[9] The operation of the mine has also had major consequences for the health of the communities living near the mine, my clients. In 2019, the Colombian Constitutional Court found that harm to human health 'will be caused or continue to be caused' by pollution from the mining company's activities, and held that 'this would imply serious and irreparable harm to the community'.[10] Among those most vulnerable, due to the location of El Provincial, were the Wayúu community, whose rights I had to protect.

To do so, I had to distil thousands of pages of evidence into twenty clean pages of submissions, compelling enough to move the UN machinery at a time when the world had come to a halt.

The operations of the mine

Sitting in my kitchen at 5 a.m., I first immersed myself in the operations of the mine.

Although the Carbones del Cerrejón concentrates its mining operation in the Middle and Low Guajira, years ago the company built a railway to transport the coal, which runs through the entire Wayúu territory. From the mine in the lower part of La Guajira, to the Bolivar port, where the coal is then shipped to the world, the 120-carriage railway operates without rest twenty-four hours a day, seven days a week. The train makes nine daily trips. Where does the coal go? I wondered.

The mine's work effectively does not stop. Open-pit mining operations mainly include drilling, blasting, mining, transportation and dumping. The blasting attracted my attention. It is incessant. The operation of the mine has an annual consumption of more than 80,000 tonnes of explosives per year.[11]

I could not imagine myself living so close to non-stop explosives activity and carbon dust. Could you? I searched for footage on YouTube in order to actually see and hear the sounds of the mine. In one of those images I saw El Provincial. I saw a road; a yellow, flat road, edged on both sides by green shrubs, which sank towards what I assumed was the horizon. When the camera got closer, you could see that what was behind was not the sky but a kind of gigantic yellow wall that had sucked up all earth and blocked any other vision. A monster. It was the mine.

El Provincial is the place where the Wayúu community that instructed me lives – it's located in the southern part of La Guajira, next to the Ranchería River. That means they are right next to the mine. The lower part of La Guajira used to be fertile; it was where the water was concentrated. It used to be the '*despensa agrícola*' of La Guajira: the place that provided agricultural products to the rest of La Guajira. But these most productive and fertile lands were given up for the extraction of coal.

The Wayúu's previously lush and biodiverse landscape, their most fertile land, 'have turned [. . .] into pure hills of sterile material, which even they call themselves [the mining company] sterile material. I hear, that woman is sterile, when she is a poor woman she cannot give birth to children. And if the land is barren, what can it give us? [. . .] How is it going to grow a Guáimaro tree there? Never.'[12]

A neighbour from hell

'I don't forget. The date the train whistled for the first time in front of my house,' recalled a woman from La Guajira. 'It was on 5 February 1985.' Like a black snake, the train slithered through. 'I can't ever forget that date. When the train arrived, the first thing it did was to destroy the cattle, the donkeys; little by little, Cerrejón took away territory. They started with the springs and then with the little towns that you can't even hear any more about.'[13]

Today, if you arrive at El Provincial, where Luz Angela Uriana, one of my clients, lives, you might first notice a 'rotten egg' smell of sulphur or 'burned coal'. You probably will not be aware of the particulate matter in the air known as PM 2.5, as it is invisible to the eye, but you'll notice that everything in El Provincial is covered in coal dust. The landscape is arid, dominated by desert with xerophytic shrubs (plants adapted to life in a dry habitat) and scrubs. From Luz Angela's humble house, you can see the open pit: Patilla. You hear the constant blasts. After a few hours, perhaps you will experience nasal and breathing discomfort, shortness of breath and stinging eyes. After a few weeks, you may experience blurred vision and headaches. The constant sound of the blasts grate on the mind and stops you from sleeping. If you stay for months, you may experience a dry cough, asthma, pneumonia, hypertension, damage to the skin and eyes. After some years you may develop cancer, damage to the cells, genetic damage, malformations and mutations in your blood structures.

The extraction and transportation of coal causes the emission of particulate matter pollutants into the air known as PM 2.5 and PM 10.* I had no previous experience with pollutants, but I learned fast that both pollutants are dangerous to human health.[14] PM 2.5, the one invisible to the eye and omnipresent in EL Provincial, is the deadliest and smallest form of particulate matter. It has been proven to cause diseases such as asthma, pneumonia, hypertension, cancer, damage to the skin and eyes, miscarriages, premature births and pre-eclampsia.[15] Its concentration in the air around Carbones del Cerrejón only began to be measured in 2018, by which time the mine had been operating for thirty-five years.[16] Thanks to pressure from the local communities, through judicial channels, the Colombian public environmental authorities finally began to measure this

* 'PM' refers to 'particulate matter'. The numbers 2.5 and 10 refer to the diameters of the respective types of pollutants in micrometres.

pollutant. The above effects were confirmed by scientific investigations.[17] Health effects for children in El Provincial were covered in a 2017 documentary, *The Curse of Coal*, produced by DW, the German state broadcaster.[18] The documentary demonstrated that the environmental impacts of the mine, particularly from air pollution, have had a severe impact on the health of children from the Wayúu communities.

Back in 2017, an action of protection (Acción de Tutela) before the Colombian Constitutional Court was filed by the indigenous women of El Provincial, represented by Rosa María Mateus from CAJAR, the Colombian collective of lawyers who first contacted me. This legal action demanded effective protection of all the human rights violated, but especially those related to the lives and health of the Wayúu children of El Provincial. Due to their greater vulnerability, children in the region have suffered serious and recurring respiratory and skin diseases, fevers, headaches and diarrhoea, and many other conditions.

The Wayúu claimants also stated that the constant roar of the noise from the machinery has prevented sleep and their ability to dream peacefully, disturbing their ancestral relationship with birth, pain and death. One member of the Wayúu community explained: 'My mother is a dreamer, and the train interrupts her dreams and she is unable to continue dreaming once she is awake. And this is upsetting because her dreams are a source of important information for us.'[19]

The Constitutional Court of Colombia delivered its judgment in December 2019, notifying the parties in February 2020.[20] It recorded that 10 per cent of the members of the Wayúu in El Provincial have affectations in their lung function and various cases of respiratory diseases, and acute respiratory tract infections were also found in this population. It was noted that high concentrations of various metals had been found in the blood of the inhabitants near the mine, especially sulphur, chromium and bromine, which can cause DNA damage and diseases such as cancer. The existence of damage in the cells of residents of the area was verified, all of which can be related to respiratory, cardiac, dermatological and cancer diseases, among others. The court ordered the mine to reduce air pollution as an 'urgent transitional measure'.[21]

However, no action was taken.

As I read the Constitutional Court Judgment, I was shocked. How could our world create such a living hell on earth?

The only river in the desert

In the meantime, the Covid-19 pandemic had worsened the position of the Wayúu. And a key challenge for the communities in the midst of the coronavirus pandemic was access to water. Over several calls, Rosa María filled me in about the unfolding situation in La Guajira.

To indigenous peoples, water is sacred. As scholars of their culture have noted, 'for the indigenous worldview, water is intimately related to the existence of the human being'.[22] In addition: 'Water for the Wayuu is a living being that is often framed within the symbolism of dreams, to reveal relational facts with the natural environment with which they interact.'[23]

The Wayúu stated that before the mine:

> [t]he soil was fertile, it allowed the grazing of goats, sheep and cows; yucca, ahuyama, bananas, corn, millet, melon [. . .] were cultivated, there was also a great variety of wild fruits. In the mornings the melodious songs of the birds could be heard, and throughout the day, the springs ran from the source to the

> mouth, along with the rain watering the green grass and the immense trees of caracolí, oak, trupillo, jobo, ceiba, guáimaro, cotoprix, mamoncillo and the algarrobillos that refreshed us and produced fruits and food for domestic and wild animals. We felt privileged with the water sources that gave us the precious liquid for our families, such as the Ranchería River, the streams, jagüeyes and springs of crystalline waters. We did not have needs, we did not have the problem of water and food that we now suffer in La Guajira.[24]

The Ranchería River was known to be the most important source of water in the department of La Guajira. It played a key role in the maintenance of ecosystems in its basin, as well as providing water for domestic, recreational, cultural, spiritual and farming activities. As of 2016, it was estimated that 450,000 people depended directly and indirectly on the water of the Ranchería River.[25] The river is sacred to the Wayúu.[26] Many people in La Guajira relied on the river for cleaning, bathing and cooking. Some also relied on the river for their drinking water.[27] However, the Carbones del Cerrejón mine is one of the largest users of water in La Guajira today. In 2019 it extracted 10,733 million litres of surface water.[28] An estimated 11 per cent of its total water extraction – 1,241 million litres of freshwater – was drawn directly from the Ranchería River.[29] These facts left me flabbergasted. The mine operation requires approximately 24 million litres of water a day, and this was in a land of water scarcity.[30] This, primarily, according to the mine, is 'for dust control on the roads where mining equipment travels'.[31]

By 2015, the shortage of water due to the operations of the Carbones del Cerrejón mine in La Guajira was denounced by the Wayúu communities from the High Guajira (Uribía, Manaure, Riohacha y and Maicao), and they requested urgent measures of protection for the risk to their lives and personal integrity caused by the 'lack of access to drinking water and the state of malnutrition that this causes to members of the community, especially girls and boys'.[32]

I had no idea, prior to taking on this case, what open-pit coal mining entailed, but as I dug into the facts, I began to learn more, and it was shocking. The mine not only uses an enormous number of streams and tributaries, but also, perhaps worse still, it returns water contaminated with heavy metals, chemicals and sediments to those same waterways.[33] In 2019 it dumped 578 million litres of liquid waste (primarily runoff from dump sites and pits).[34] Research has shown that manganese, selenium, barium and strontium are all present in higher concentrations close to where the mine dumps its waste materials.[35]

A 2017 analysis found that, as a result of the mine, various metals known to cause serious health effects were present in the waters in and around the Ranchería River.[36] Specifically, it found that the levels of lead, cadmium, barium, manganese, iron and zinc surpassed permissible levels under World Health Organisation (WHO) guidelines. A subsequent study published in July 2019 also found dangerously high levels of mercury in the water.[37] Long-term overexposure to these metals, particularly in drinking water, causes nausea and vomiting,[38] impaired kidney function,[39] constriction of blood vessels,[40] muscle pain and muscle weakness,[41] neurological disorders, and – in cases of particularly high or long-term exposure – death.[42] Mercury, which appears in the WHO's top ten chemicals of major public health concern, can have toxic effects on the nervous, digestive and immune systems, and on the lungs, kidneys, skin and eyes, even in small quantities.[43]

In December 2019, the Constitutional Court of Colombia found that:

> [. . .] the surface and underground water sources of [the Provincial] community were being affected by the Cerrejón operations, due to the contribution of contaminating sediments and the disappearance and alteration of channels and aquifers.
>
> Non-compliance with the discharge regulations [was identified] and the presence of oily liquid residues from the company, as well as coal-like material were found in the Ranchería River. In addition, discharges that were carried out without the corresponding permission were evidenced.[44]

The Constitutional Court ordered Carbones del Cerrejón to 'prevent contamination of nearby water sources' as an urgent matter.[45] But, yet again, the order was not complied with. Instead, Carbones del Cerrejón criticised the court's ruling in its 2019 Sustainability Report.[46] This report also made repeated references to the fact that Carbones del Cerrejón provides drinking water to affected communities.[47] However, as several non-governmental organisations (NGOs) monitoring the situation in La Guajira have pointed out, potable water has been provided to *some* of the affected communities, but not all. Besides, the 450,000 people who live in the region did not need drinking water to be provided for them prior to the mine's operation. It was a piecemeal solution to a problem the mine itself had created.[48]

The community that reached out to me no longer has free access to water. Instead of access to clean water from the main rivers and tributaries, as had once been the case, water was periodically delivered via tanks on the back of trucks. Luz Angela Uriana Epiayú, a Wayúu woman from the affected community whose children have contracted diseases as a consequence of the contamination by the mine, is leading the El Provincial community's fight against

Carbones del Cerrejón. She put it succinctly: 'the water they bring by tank car to some communities is a bit of the water that they have stolen from us and polluted.' [49]

Is there a right to water?

You may be wondering is there an entitlement to water? According to international law, yes. Water is an essential human right. On 28 July 2010, the United Nations, through its General Assembly, issued Resolution 64/292, entitled 'The human right to water and sanitation', which recognises that 'the right to drinking water and sanitation [. . .] is an essential human right for the full enjoyment of life and all human rights.'

In the case of indigenous peoples, who are tied to a particular ancestral land, 'access to water resources on their ancestral lands is protected from encroachment and unlawful pollution. States should provide resources for indigenous peoples to design, deliver and control their access to water.'[50] The International Covenant on Economic Social and Cultural rights, a binding and nearly universal treaty, provides that a people (and the Wayúu are, for the purposes of the law, a people) may not be deprived of 'its own means of subsistence'. The committee in charge of the interpretation of such an instrument has stated that 'States parties should ensure sufficient access to water for subsistence agriculture and to ensure that of subsistence of indigenous peoples.'[51] These, in effect, were the rights of the Wayúu.

Beyond their right to water, the Wayúu predicament was so egregious to me because the law in the Americas had already recognised the right to a healthy environment as 'a fundamental right for the existence of humanity', and part of the right to life.[52] The Inter-American Court had upheld that the right to a healthy environment is a right with individual and collective connotations, and that environmental degradation affects the effective enjoyment of human rights (including, fundamentally, the right to life).[53] This regional court (the Americas equivalent of the European Court of Human Rights) has further held that 'several rights may be affected as a result of environmental problems, and that this may be felt with great intensity in certain groups in vulnerable situation' among which indigenous peoples and 'the communities that, essentially, depend economically or for their survival, fundamentally on environmental resources, [such as] from the forested areas and river basins'[54] are found.

If all of this was the law, how, then, did Carbones del Cerrejón (i.e. Anglo American, BHP Billiton and Glencore) get away with the pollution of the Wayúu's ancestral land?

They relied on the Wayúu being unable to enforce those laws. When the Wayúu managed to get Colombian courts to rule in their favour, the mine simply didn't comply. The country's institutions were not strong enough to deal with some of the most powerful fossil fuel companies on the planet.

The diversion of waterways

When I first came across the Constitutional Court's reference to 'the disappearance and alteration of channels and aquifers', I didn't quite understand. Then it hit me. Something deeply disturbing had been taking place.

The mine was not only polluting the existing water sources, but it had rerouted altogether an enormous number of streams and tributaries to mine coal from beneath them. The flow of the Ranchería River used to be fed by twenty-three main tributaries, for example, and a large number of streams, many of which have been destroyed, disappeared and diverted to allow the expansion of the mine.[55] It is estimated that Carbones del Cerrejón has diverted more than seventeen streamlets[56] and damaged another thirty streams[57] in the region. In fact, its activities have caused irreparable damage to the hydrographic basin of the Ranchería River.[58] The deviations also affected the underground water, which, experts note, 'is the most affected by the mining'.[59] And the damage goes beyond the river itself. 'The Ranchería River is a natural retainer of the biodiversity of Sierra Nevada. Without the river, the Sierra will gradually become desertified.'[60]

In 2016, Carbones del Cerrejón decided to divert the Arroyo Bruno, a Ranchería tributary,[61] to enable the extraction of approximately 35 million tonnes of coal from beneath the riverbed.[62] To the Wayúu, for whom streams and waterways are sacred, this is sacrilege. You don't touch the course of a river. When reading about all this, I imagined the reaction of Londoners if the Thames were to be diverted for the interests of some private company. There'd be uproar. But despite a Constitutional Court decision in favour of communities seeking a halt to this diversion, the diversion of the Arroyo Bruno remains in place due to lack of compliance with Constitutional Court rulings.[63]

While reading these reports, immersed in the reality of La Guajira from my flat in London during lockdown, I felt indignation. I wondered on what basis could any private entity dare to play God, carrying out structural changes and interventions to the waterways of a country, all in the name of greed: the exploitation of coal. In the best Latin American legal theory tradition, broadly imbued in law faculties across the continent, the deviation of streams and rivers to extract coal from underneath could only be regarded as against natural law. A monstrosity.

And how was the mine allowed to build a pit 100 metres away from a community with children? There had been no consultation.

The mine had just expanded, eating up more and more land belonging to the Wayúu.

Displacement

For the mine to expand, many Wayúu communities have been displaced. Afro-Colombian and campesinos communities, who also live in the region, have faced forced displacement too. At times, evictions have been carried out with armed guards, tear gas and metal projectiles. In 2016, bulldozers were used to destroy an afro-Colombian village. Even when Carbones del Cerrejón claims to have consulted with displaced communities, NGOs representing these communities attest to the fact that it has not offered a genuinely free choice as to relocation. 'Consultation' is premised on the assumption that expansion will continue.

Displacement was not a choice for the Wayúu community in El Provincial. My clients were living in their ancestral land. It is simply not an option to move away. 'This is the land of our ancestors, the heritage of my children,' as one of my clients put it at a meeting I organised. International law recognises collective rights of property and possession to indigenous peoples upon their ancestral land tied to their right to dignity. The United Nations Declaration on the Rights of Indigenous Peoples, in Article 25, states:

> Indigenous peoples have the right to maintain and strengthen their distinctive spiritual relationship with their traditionally owned or otherwise occupied and used lands, territories, waters and coastal seas and other resources and to uphold their responsibilities to future generations in this regard.

Article 26 of the same Declaration, meanwhile, states:

1. Indigenous peoples have the right to the lands, territories and resources which they have traditionally owned, occupied or otherwise used or acquired.
2. Indigenous peoples have the right to own, use, develop and control the lands, territories and resources that they possess by reason of traditional ownership or other traditional occupation or use, as well as those which they have otherwise acquired.
3. States shall give legal recognition and protection to these lands, territories and resources. Such recognition shall be conducted with due respect to the customs, traditions and land tenure systems of the indigenous peoples concerned.

The law says that the government can't allow a mine to be built if the people in the local area aren't consulted. It requires a state to obtain the free, prior and informed consent of the indigenous people concerned prior to a major development project.[64] A consultation has to take in their concerns, and explain to them the realistic likelihood the mine will have on them and their way of life. Good faith consultations imply a prior environmental and social impact assessment conducted with indigenous participation, and reasonable benefit sharing.[65] The purpose of the environmental and social impact assessment ensures that the indigenous people are aware of possible risks, including environmental and health risks.[66] In addition, the environmental and social impact assessment should address the cumulative impact of existing and proposed projects.[67] For example, the cumulative impact on an expanding coal mine on the environment and on the human rights of the Wayúu would have had to be studied prior to approving an expansion of the mine.

Projects that affect the survival of indigenous groups are not legal under international law.

And the expansion of the Carbones de Cerrejón mine had, in my view, reached the threshold of threatening the survival of the Wayúu.

'The notion of "survival" [referred to in the instances for non-approval of projects], is not tantamount to mere physical existence.'[68] 'Survival' of an indigenous or tribal group 'must be understood as the ability of the people to preserve, protect and guarantee the special relationship that [they] have with their territory, so that they may continue living their traditional way of life, and that their distinct cultural identity, social structure, economic system, customs, beliefs and traditions are respected, guaranteed and protected.'[69]

In the case of the Wayúu, free, prior and informed consent had not taken place before the mine's expansion. No impact assessments had *ever* taken place. The Wayúu did not benefit in any way whatsoever from the destruction of their natural surroundings and health. My clients were correct to be asserting their rights to their ancestral lands and to their cultural connection with it. In fact, they were fighting simply for their survival.

The 2019 judgment by the Constitutional Court had been only the latest in a series of judicial findings denouncing Carbones del Cerrejón's activities. The Constitutional Court had identified the risks involved in mining in La Guajira as early as 1992,[70] and the mine had been the subject of litigation many times.[71] As of 2025, Carbones del Cerrejón continues to operate under a licence granted in 1983,[72] arguing that this licence 'exempts it' from complying with current Colombian environmental law – an interpretation that has been rejected by the courts.[73] In 2017, the court reviewed a vast body of academic literature on the effects of the Carbones del Cerrejón mine, and summarised the effects of the mine on the fundamental rights to health, water and food sovereignty.[74] It also held that Carbones del Cerrejón had given insufficient consideration to social and environmental impacts when deciding to expand the mine.[75]

La Yonna: The sound of a kasha

It was the women of El Provincial who decided to file a lawsuit beyond the jurisdiction of Colombia. The women did so because of their children. Within the indigenous communities, it was children who were the most vulnerable. I had seen this in further evidence. Rosa María forwarded pictures documenting the skin diseases of the children of El Provincial as a result of the mine's pollution, and information about their condition.

Some male traditional authorities of the El Provincial community had opposed bringing a claim against Carbones del Cerrejón. But this did not stop the women. The women found strength because of their children. Wayúu women, after all, are the fundamental axis of the Wayúu culture. As Carmen Uriana, one of the women of EL Provincial, said:

> We are victims. A little girl of mine had her little lung burst. We don't have our water to be healthy, we have a little river, but every now and then they are threatening that they are going to divert it because it has coal beneath. We are like children when they are hungry and cry, we are asking to be seen.[76]

Carmen Uriana's testimony was backed by evidence. In addition to the pollution, the lack of water had destroyed the ecosystems and created food scarcity in the Wayúu's ancestral land. Between 2016 and 2018, an average of one indigenous child under five had died every week in La Guajira because of malnutrition.[77] An unimaginable loss, every week. The high level of infant mortality among the Wayúu people was highlighted by the Inter-American Commission on Human Rights (IACHR), a regional body headquartered in Washington, DC, in 2015, when it directed the Colombian government to take immediate precautionary measures to safeguard the lives and personal safety of the Wayúu people in La Guajira.[78] The IACHR's decision was prompted by the documented deaths of 4,770 Wayúu children during the preceding eight years as a result of thirst, malnutrition and preventable disease.[79] This word lingered in my brain – *preventable*.

The Wayúu women of El Provincial had won my respect. In their determination I could hear the sound of a *kasha*, a small drum used in the traditional ritual Wayúu dance called La Yonna. On the golden sand, barefoot, Wayúu women dressed in red *sheins* (long loose-fitting dresses made of cotton), the red colour symbolising life and vitality, dance to the beat of a *kasha*, imitating the movements of certain animals considered to be strong and fast. The women dance as if gliding through the sand. Their arms are extended like a bird and their hair covered with a red veil. It is a magnificent sight. I don't know the full symbology, yet I can feel its strength. The women wear Wayúu makeup, called *Acheepa* – spiral symbols and butterfly wings can be seen.

The women go from right to left, forward and backward, diagonally and zigzagging, just as a mimesis of birds and mammals that inhabit the Guajira's geography might. Often, the men dancing with them fall. I listened to La Yonna while drafting their complaint. It gave me strength to work long hours and persist, to channel their own persistence. They dance it as an affirmation. It is a dance affirming life and representing the articulation between humans and nature with which they honour their ancestors.

The resistance of the Guáimaro tree

Like the Guáimaro tree, the women were resisting. This tree grows in the dry ecosystem of La Guajira. Thanks to its roots, which extend up to fifty metres deep into the earth, it is resistant to droughts and hurricanes. It can live for a hundred years and is productive until its death.[80]

The Guáimaro tree is sacred for the Wayúus. It bears fruits and nuts, and is more densely nutritious than even the avocado.[81] In fact, it has as much protein as milk, four times more potassium than bananas, as much iron as spinach and four times the magnesium of red kidney beans. The tree was revered by the Wayúu for its medicinal properties, capable of treating respiratory diseases and rheumatism.[82] But, due to the environmental impacts of the Carbones del Cerrejón mine, they can no longer engage in cultural and commercial practices associated with this tree.[83] The trees here are dying. Their increased disappearance as a direct result of the expansion of the mine has led to both food insecurity and a loss of cultural heritage in La Guajira.[84]

Other surviving flora in the area has also been harmed by pollution generated by Carbones del Cerrejón's operations: in a December 2019 judgment, the Constitutional Court found that 'the flora of the region [has been] affected by its proximity to the mining complex, causing constant exposure to pollution and the accumulation of particle material.'[85] The loss of indigenous plants has impaired the practice of customary traditional medicine by the people of La Guajira.[86]

The Wayúu's customs include a set of principles, procedures and rites that govern the social and spiritual conduct of the community. This system has been inscribed in the List of Intangible Cultural Heritage of Humanity of UNESCO.[87] Yet the environmental devastation of La Guajira, and the forcible displacement of its people by the mine, has led to irreparable cultural harm. The cultural heritage of the Wayúu is inextricably linked to their ancestral lands.[88] As one member of the displaced Tabaco community explains:

> We ethnic communities, Afro-descendant and Wayúu, have always lived off of agriculture, fishing, hunting, and from herding our animals. We have a spiritual anchor to our land [. . .] Because we have been displaced, we have lost our sacred places, our meeting places, we have lost our ancestral medicine.[89]

As well as devastating the natural environment, the mine's expansion has led to the destruction of indigenous communities' churches and cemeteries.[90] The bulldozing of sacred sites is a direct incursion into the cultural heritage of the Wayúu people.

One of my clients from El Provincial, Cristina Epiayú, an Ouutsü woman leading the fight against Carbones de Cerrejón, stated:

> I know traditional medicine, but as the multinational is ending the health of our children, so is traditional medicine ending, because there was a sacred mountain where our grandparents went to look for plants, there are no more and if there are, they are few and they cannot be given to children because they are contaminated, dirty with coal dust. With the vibrations [of the mine] we no longer dream as we dreamed before, we no longer sleep well because the noise sounds every night.[91]

The damage caused to the cultural property and identity of indigenous people in La Guajira – a region where almost half of the population is indigenous[92] – is both immeasurable and irremediable.

This was one of the most devastating cases I had ever come across. It was the sort of humanitarian crisis, however, that did not make any headlines, a type of crisis that was invisible to us all in the West.

But that needed to change. Despite all they had been through, the women of El Provincial had decided to continue their legal fight. They were resilient like the Guáimaro tree, whose roots grow very deep.

A lot of logistics had taken place to organise a virtual meeting early in June 2020. Luz Marcela from CAJAR had managed to arrange things so that I could be connected from my living room, still in lockdown in London, to La Guajira. My clients, however, had to travel some distance to be able to reach an internet connection. There were a few of them together, including a female traditional authority, an Ouutsü, and also some male members of the community. They were sitting in a circle, connecting from the open backyard in a *ranchería*, surrounded by shrubs, outside a house. I could see the vastness of their open sky. I asked first what they wanted to achieve by way of an immediate measure. One of the elders in the group answered with no hesitation: 'The closure of Tajo Patilla' [the pit next to El Provincial], and the suspension of the mine work during the Covid-19 crisis.'

By 2019, the Paris Agreement, the main treaty on climate obligations, had made it necessary for states to start phasing out fossil fuel projects. So the Wayúu's objective was not fanciful, but required by the new international regime. Closing the mine, however, was going to take more than an urgent request. A legal strategy had to be put in place to ensure that. Within that longer-term strategy, a request to suspend the operations of the Patilla pit during the Covid-19 crisis was particularly urgent and essential for their survival. The Wayúu also wanted the world to accept that a mine causing the harm I have described above, in their ancestral land, their homes, was wrong. With this in mind, we had to start by approaching somewhere that could provide a prompt response. Unlike cases where the facts hadn't previously been established, the facts of this case had been proven before the highest court in Colombia. Therefore, I prepared the claim to be filed before the United Nations Special Procedures complaint mechanism.

The UN Special Procedures are Geneva-based mechanisms established by the Human Rights Council to report and advise on human rights, from a thematic and country-specific perspective. Independent experts are appointed for those purposes and given specific mandates. For example, they have a Special Rapporteur for 'safe drinking water'. The complaint mechanisms under these procedures are not based on a treaty. Rather, they emanate from the Human Rights Council's powers, which are based on the UN Charter. These complaint mechanisms listen to all sides of a particular case referred, look into the facts of the matter, assess whether or not there have been human rights violations, and issue recommendations. These Special Procedures can look not only at state actions but also at actions by corporations. This was of particular importance for this case. There are expedited procedures when there is an emergency. This was the procedure that I seized upon.

I made an urgent request for the intervention of the Special Rapporteurs (experts appointed to these mandates) on the issues of human rights obligations relating to the enjoyment of a safe, clean, healthy and sustainable environment; of human rights to safe drinking water and sanitation; and of the right to food; and the Working Group on the issue of human rights and transnational corporations and other business enterprises. I knew that moving the UN machinery during lockdown when everyone was working remotely was not going to be easy. But I had to try my utmost.

In the process of working on this case something deeply disturbing came to light. It was not comfortable to reckon with. I had seen, at first hand, the real price of coal mining. What became unbearable, though, was to learn how interconnected places like La Guajira are with us, here, in our homes in cities like London.

I had been wondering where the coal went. In disbelief, I eventually found that the coal was shipped to Europe. Carbones del Cerrejón was owned by three of the most powerful multinational companies in the world, two of which are registered on the London Stock Exchange. They have an office in Ireland, from where the coal was sold to Europe and other parts of the world. They produced coal and ruthlessly expanded because there was such a high demand for coal in Europe. In December 1999, the UN Committee on the Elimination of Racial Discrimination examined Ireland's compliance reports under this treaty.[93] Observing that 'the operation of the Cerrejón mine complex in La Guajira, Colombia' [. . .] has been linked to 'serious abuse of human rights affecting the indigenous population', and noting that Ireland had purchased coal for one of its power stations in County Clare, it recommended Ireland to 'consider stopping purchasing coal from the Cerrejón mine complex'.[94]

But Ireland had not been the only European country buying coal from Carbones del Cerrejón.

For example, Carbones del Cerrejón sold coal to Turkey, and also to Germany in 2015, 2016 and 2017.[95] Germany was by far the biggest buyer of Cerrejón coal in Europe.[96] I had lived in Germany for eight months. Maybe my computer had been powered with electricity produced with that coal. But there was more.

The three giant UK-listed mining companies, BHP Billiton, Anglo American and Glencore, received billions of pounds in finance from UK banks and pension funds.[97] NGOs 'following the money' asserted that HSBC,[98] Barclays, RBS and Lloyds TSB all financed the companies behind the mine.[99] Research by Global Justice Now indicated that, between 2009 and 2013, HSBC had given £3.133 billion, Barclays £3.524 billion, RBS £3.340 billion and Lloyds £3.429 billion.[100] This made me rather uncomfortable. I myself banked with one of these banks.

All of a sudden, I felt complicit.

The Wayúu's complaint goes global

I filed the claim on 17 June 2020, and planned a strong media strategy as part of my advocacy. Climate News Hub released a short video on the case, and the next morning the news was reported by Reuters and the *Sydney Morning Herald*. In addition to working on the case, I gave interviews in different time zones and accepted requests for quotes. The Wayúus' complaint was going global.

The outgoing UN Special Rapporteur for toxic waste saw the news reporting my filing of the case. He contacted me. He had authored a report back in 2019 on the State's duty to prevent exposure to toxic waste.[101] He was preparing his last report for the Human Rights Council before leaving office, which focused on Covid-19 and States' duties. As I read it later, I saw that he noted in his report that 'The COVID-19 pandemic has increasingly illustrated the importance of safeguarding a healthy environment [. . .] as a human right.'[102]

During our conversation he said something that stuck in my mind: 'Although the UN Special Procedures do not have a way to enforce recommendations, findings of the Special

Rapporteurships can nevertheless have a powerful effect.' He told me the story of a company that they had found in breach of international law. He had watched how, the very day the UN released its findings, the company's shares on the stock market had dropped in value. Carbones del Cerrejón could ignore the rulings of local courts, but they would feel the effect if investors pulled their support.

As part of the procedure before the UN, I sought an online meeting between the Office of the Special Rapporteur on Environment and Human Rights and the Wayúu. I thought that they had the right to be heard, to express with their own words what they were experiencing. It was a historic meeting: the first time the Wayúus had addressed an office of the UN, all the way from La Guajira. They were dressed in their traditional clothing (the women were wearing their *shein* Wayúu – a long, loose-fitting dress decorated with intricate geometric patterns, flowers and animals – and headscarves called *pannera*). They were in the same place where we had spoken during our initial meeting, outdoors. Following my oral submissions, a female traditional authority opened their intervention in their native language, and then continued in Spanish. She said: 'The Carbones de Cerrejón mine has appropriated our territory. It does not respect the rights of the Wayúu people. It has contaminated our sacred rivers; it is killing them. It has poisoned the earth. You can't breathe. The earth is sick. Our children die from pollution. We want justice. We don't want the mine in our home.'

Among those testifying were the Ouutsü women, the women blessed with the special skills of dreaming.

The UN officer participating in the meeting on behalf of the Special Rapporteur's office listened carefully. She had stopped making notes. She looked at the Wayúu women intensely, as if trying to grasp the gravity of what they were recounting.

I remained silent at that point. I wanted the Wayúus to speak directly, with their own voice. There was a palpable sense that this moment was historic.

After the meeting ended, the complaint was forwarded to the other parties. Both the state and the mining company had to be given the opportunity to reply to the claim made. We waited for weeks.

Then, *bang* – on 18 August 2020, I saw the news that BHP had announced plans to sell off its thermal coal mines within two years. Was this the result of the embarrassing reports of its role in coal mining in Colombia? Of our complaint going viral? The move, reportedly, was to ready itself for a low-carbon future following consistent pressure from investors, including Norway's sovereign wealth fund, which 'owns more than US$2bn in BHP shares'.[103] It was pointed out that the high cost of remediation associated with the Carbones del Cerrejón mine '[might] make them difficult to sell to a third party'.[104] Yet they were leaving Carbones de Cerrejón. Making the violations against the Wayúu by Carbones de Cerrejón visible to the shareholders, to the world, had been crucial.

Just a month later, on 28 September 2020, the UN Special Rapporteurships pronounced themselves in favour of halting mine operations next to the El Provincial reserve until it could be made safe. This was unprecedented. In effect, the UN Special Rapporteurships were calling for the coal mine to suspend its operations.

Special Rapporteur David Boyd, an associate professor of law, policy and sustainability at the University of British Columbia, who has advised many governments on environmental, constitutional and human rights policy,[105] remarked: '[t]he situation that was brought to my attention recently regarding the El Cerrejón mine and the Wayúu indigenous people, is one of the most disturbing situations I have learned about in my two and half years as Special Rapporteur'.[106]

The following statement was released by the UN:[107]

> 'I call on Colombia to implement the directives of its own Constitutional Court and to do more to protect the very vulnerable Wayúu community on the Provincial indigenous reserve against pollution from the huge El Cerrejón mine and from COVID-19,' said David Boyd, UN Special Rapporteur on human rights and the environment. 'At least during the pandemic, operations at the Tajo Patilla site close to the Provincial reserve should be suspended until it can be shown to be safe.
>
> 'It is absolutely vital that Colombia protect the indigenous peoples' rights to life, health, water, sanitation, and a safe, clean, healthy and sustainable environment by halting mining close to the Provincial reserve until it can be made safe.
>
> 'I further call on the mining company to increase its effort to prevent further harm to people and also to ensure that those who have been negatively impacted have access to effective remedy.'

Special Rapporteur David Boyd's call was supported by six other UN special rapporteurships, and the Working Group on the issue of human rights and transnational corporations and other business enterprises.

The Wayúu women had won. The next day, the Wayúu of El Provincial sent a recording celebrating their victory. They were standing in front of the mine, in traditional clothing, offering a salutation to the UN Special Rapporteurships for upholding their plight.

A first victory

It was clear to me that the complex situation in La Guajira would need more than one UN intervention in order to be resolved, but the

UN acknowledgement meant a great deal to the Wayúu. The world had seen and accepted that they were being wronged. The reaction of the mining company to the pronouncement of the UN was, unsurprisingly, not positive. The news had reached them in the midst of the longest coal strike in their history, from 31 August 2020 until 1 December 2020.

Encouraged by our success before the UN, several NGOs filed a claim under the Organisation for Economic Co-operation and Development (OECD)'s Guidelines, a set of rules for responsible business conduct, directly against the parent companies BHP, Anglo American and Glencore for lack of compliance with human rights and environmental rights. The OECD has established complaint processes by which parties (i.e. a community affected by the companies' activities or NGOs) can submit complaints against multinationals in breach of the guidelines, and I was instructed to draft and advise on this complaint as well.

The complaint demanded the closure of the Carbones del Cerrejón coal project for good because it was responsible for 'serious human rights abuses and devastating environmental pollution'. This time the complaint did not focus on one community only but covered many communities across La Guajira. The measures requested included the mining company having to have an exit plan in La Guajira. We filed the complaint in January 2021. This time the news of the complaint made it to the *Financial Times*.[108] Shortly after the complaint was filed, Anglo American and BHP exited Carbones del Cerrejón, leaving Glencore as sole owner of the mine.

On 10 January 2022, the Swiss, Australian and UK National Central Points (NCPs) (i.e. the bodies that receive complaints under the OECD procedures) issued their initial assessment under the OECD Guidelines in relation to the three international mining giants (BHP, Anglo American and Glencore) over serious human rights abuses relating to environmental pollution at the Carbones del Cerrejón coal mine. They found that the complaints dealt with a credible issue and should progress. A mediation process to ensure the companies provide remediation was begun.

The legal fight of the Wayúu women continues, but the first step, which internationalised their plight, was won because of their courage and determination.

Soon after filing the UN claim, I happened to attend a Zoom meeting in which lawyers working on environmental matters across Latin America were giving presentations. I realised, all of a sudden, that they were all women – highly represented in environmental NGOs. Then the penny dropped. It was a moment in which I realised that women – the Wayúu women, the CAJAR lawyers, the many women working for the NGOs that brought the complaint before the OECD, and myself – were leading the way as defenders and guardians of the natural world.

Still, the final closure of the mine remains a pending task. It is urgent. Our life in the West should not be built on pain and suffering, on the destruction of the lives of others, on the existence of living hells. It cannot be justified to run electricity on the back of the suffering of indigenous communities like the Wayúu, whose land is contaminated and who stand today dispossessed of the most essential element for survival: water.

But just when I thought we were finally getting rid of such wrongs, I had a telephone conversation with Rosa María in 2022, when the Ukrainian war was ongoing. Sanctions against Russia had caused Europe to search for energy from elsewhere. Russia used to sell coal to countries like Germany. Not able to buy coal from Russia, the Germans turned back to Colombia for coal . . . coal from the Carbones del Cerrejón mine.

TWO

Myanmar: Southeast Asia's Last Free-flowing River

My connection to rivers began with the Peruvian Amazon. When I'd first visited, back in 1984 and 1985, I hadn't quite grasped that I was in paradise. The rainforest was still relatively pristine and lush. The Spanish conquerors had been unable to penetrate the jungle and, although some illegal logging had already taken place, oil companies had not yet set foot in it at that point either.

During this visit, I was among people – Huitotos, Amueshas, Ashaninkas, Boras – for whom post-colonial boundaries meant nothing. Peoples who spoke their own language and not the language of a coloniser. They had their own ways and lived in harmony with the jungle. It was refreshing to acknowledge the beauty of their natural state.

What is embedded in my memory is the dark chocolate colour of the Amazon River. The torrential rain. The travelling upstream – for days – until we came upon communities where children were running around naked and swimming in a natural pool formed by the end of a stream. After days of being in a peque-peque raft in the tropical heat, I swam with them, even in streams said to be inhabited by pirañas. 'Not to worry – there is plenty of food for them here,' I was told by the raftsman.

I was reminded of these scenes many years later, when I visited Teotihuacan in Mexico and saw a pre-colonial fresco called 'The Paradise of Tlaloc', depicting aboriginal Mexicans running around in a state of nature, hunting butterflies with nets and swimming in rivers and lakes. It brought to mind the children of the Bora community, enjoying water that was uncontaminated, existing in a sort of paradise in the middle of the Amazonian jungle. The intensity of the sun and the state of bliss – that remains within me.

Meeting Asia's last free-flowing river

Have you ever felt compassion for a non-sentient thing?

Rivers, mountains, volcanos, certain rock formations, of course, are regarded as sentient by indigenous cultures. Nature and its component streams, lakes, rivers, trees and rocks are regarded as 'our kin'.[1] This notion is in my DNA. My direct ancestors, the Cavanas, believed they emerged from a snow-capped mountain in front of the town of Cabanaconde, where my grandparents and parents were born.[2] It's an extinct volcano over 6,000 metres high, called Hualca Hualca. In the Andes, mountains have genders.[3] The Hualca Hualca is female, and she is sacred. It is not difficult to understand why. Its snow-melt water irrigates the moss-green terraced fields on which this agricultural community depends. These beautiful pre-Inca fields produce the most wonderful corn I have ever tasted. Known as *maiz cabanita*, this corn is in my blood. It only grows in Cabanaconde. I have a vivid mental image of the riot of colours, the deep red, yellow, orange and speckled purple of this corn, and the way it was dried after harvesting, up in the barn on top of the entrance to my grandparents' house on a sunny morning. The Cavanas likened the waters of the Hualca Hualca to a mother's milk. They consider themselves to be children of the Hualca Hualca. That is my origin. I, too, am a child of the Hualca Hualca.

Perhaps, because of my roots, I find it easy to understand how a river could arouse compassion. The first time I experienced this feeling for myself was during one of my first instructions after qualifying as a barrister in my practice in London. The case concerned anticipated hydropower developments along the Salween River in eastern Myanmar (or Burma, as it was formerly known). This case, which on the face of it appeared to me a mere technical piece of advice, put me in touch with the ecology of a river and with peoples living in this remote area of Asia for the first time. And as I explored their story more deeply, I experienced compassion not only for the predicament of the riverine communities depending on the river, but for the river itself.

The Salween River is one of the longest free-flowing rivers in the world. It's over 3,000 km long, flowing south from the Tibetan Plateau, through parts of China,[4] Myanmar and Thailand, and finally into the Andaman Sea. Known as the 'Grand Canyon of the East', the river cuts deep into the earth[5] and its basin supports some of the world's most biologically diverse ecosystems.[6] The river is home to over 7,000 species of plants and eighty species of rare or endangered fish. From its headwaters in Tibet to its estuary in Myanmar, it remains relatively untouched. Flowing through remote and pristine regions, the river supports the livelihoods of over 10 million people, sustaining the rich fisheries and fertile farmland crucial to the survival of indigenous and ethnic minority communities living along its banks.[7]

It is the longest undammed river in mainland Southeast Asia.[8] However, this undammed status was threatened by *seven* proposed dams: the Kunlong, the Nong Pha, the Mong Ton/Tasang, the Ywathit, the Weigyi, the Dagwin and finally the Hatgyl. The largest of these was the Mong Ton at 7,100 megawatts (a megawatt being a million watts), with a basin area bigger than Singapore.

The scale of it left me gasping. The disruption and biodiversity loss because of such a gigantic project is unimaginable. As I began my research into this area, I learned that the Mong Ton forests were home to rare species, such as the Indochinese tiger, the clouded leopard, the Sunda pangolin and hundreds of other globally important species. Decades of armed conflict and the remoteness of the Salween mountains and valleys had left the river and its ecosystems mostly understudied. But the Mong Ton project's vast reservoir would inundate unexplored worlds of global ecological importance, as well as unleashing myriad other environmental and social consequences for the place, its people and biodiversity.[9] An entire segment of wilderness would be wiped out if the project were to go ahead.

The naturalist Edward O. Wilson has said that 'history is not a prerogative of the human species. In the living world there are millions of histories. Each species is the inheritor of an ancient lineage'[10] and the history of each species can be considered 'an epic'.[11] The Salween River was a living whole. It carried life and was part of a complex and fragile ecosystem. It felt appallingly wrong to murder such a life.

In 'Do Dams Violate a River's Right to Flow?', Darlene Lee posits that dams alter a river's ecosystem, 'with devastating consequences for wildlife'.[12] Species extinction is one of the identified adverse effects of dams. A dam blocks the free flow of a river. It changes its temperature, chemical composition. Its waters become stagnant and deoxygenated. The river becomes unsuitable as a habitat for the animals (in particular fish) and aquatic plants that have evolved within it.[13] Lee notes that, becoming aware of this in recent times, many countries have started to decommission dams.[14] This is quite right. In 2023, I watched recorded material of such decommissioning – the case of the Klamath River dams in the US, for example, said to be the largest dam demolition in history. 'Some of these projects have a significant impact on the environment and a significant impact on fish,' according to an authority involved in the decommissioning of the Klamath River dams.[15] The removal of the dams restored the Klamath, the second-largest river in California, to a free-flowing state for the first time in more than a century. Behind this step was the Yurok, a Native American tribe. Salmon from the river had sustained the Yurok 'since the beginning of time'.[16] The removal of the dam meant that the Klamath salmon could finally, as they put it, come home.

Back in 2016, in my office in London, I was researching the specific situation of the Salween. Nang War Nu, an ex-member of parliament for the Shan Nationalities Democratic Party, explained the world depending on the Salween as follows:

> In Kunhing township, and throughout the Pang and Salween watersheds, rural populations depend on the river and the forest to survive. They grow rice and vegetables in the lands surrounding the river, their lowland farming tracts dependent on river water for irrigation. The river is harvested for fish, crab, prawns, snails and oysters. The surrounding forests provide vegetables and mushrooms, traditional medicines and firewood. For the moment, some communities here get their electricity from mini-hydropower on the Salween's tributaries. Kunhing township is dotted with historical sites of deep significance to Shan people, including ancient pagodas, traditional Shan chiefs' houses and sacred cave temples.[17]

Not surprisingly, the dams in the Salween, to be built mainly with Chinese and Thai investment, faced widespread opposition. The communities I was to meet alleged that they '[were] being planned in complete secrecy, with no participation from affected communities and no analysis of the cumulative impacts or seismic risks of these projects.'[18] An estimated 90 per cent of the electricity generated by this project was to go to China and Thailand.[19] As groups opposing the project in Myanmar put it, 'the electricity from the projects will be sold to foreign countries, so obviously would benefit neither this country nor its people.'[20]

My involvement in this story came through the advisory instruction to assist a civil society organisation called the Karen Environmental and Social Action Network (KESAN). KESAN was a community-based non-governmental organisation (NGO) in the Karen State of Myanmar. I had to advise KESAN on the legal framework relevant to any planned hydropower development and on the information that governmental and the private sector should be sharing with the communities, to ensure that their interests were respected and taken into account in the development process. This included identifying any environmental and human rights violations in the manner in which the development process was unfolding. Effectively, my work meant giving a voice to the local communities via the language of the law.

I was informed by the communities that the plans to build the dams on the Salween River had been started by the previous military regime, but had been accelerated by Thein Sein's government. In September 2014, the Deputy Minister of Electrical Power (DMEP), Maw Thar Htwe, had announced during a parliamentary session that the government would forge ahead with the Salween dam projects, despite heavy local opposition. Then, on 2 February 2016, without informing or consulting Myanmar's citizens, they signed an agreement with China allowing implementation of eighteen out of twenty-nine hydropower dam projects planned by Chinese companies in Myanmar.

At the time, Myanmar was going through a Peace Process, meaning a formal transitional process from a dictatorship, into a democratically elected administration. A Nationwide Ceasefire Agreement had been signed in 2015, promoting long-term peace negotiations between the armed ethnic minorities and the government. The November 2015 general elections in Myanmar had also secured the victory of the National League for Democracy (NLD), which was to commence its term on 1 April 2016, ending fifty-four years of military rule. Part of my role would be to advise KESAN in meetings and directly engage on its behalf with the relevant officers of the upcoming government, the NLD, to influence policy making.

After a number of preparatory meetings working with the legal team, virtually, from New York, Myanmar and London, I found myself travelling to Yangon in Myanmar. An American lawyer with a corporate background and expertise in crisis management in riverine and estuarine matters in the energy sector had also joined the legal team. I would be advising on corporate due diligence practices under UN Guiding Principles on Business and Human Rights, as well as the rights of indigenous local communities under international law, including possible collective rights to territory and natural resources, right to consultation, and the aspects that Environmental Impact Assessments ought to address under international law. Given the likely effects of the dam and the lack of consultation on the likely affected communities, these communities wanted to stop the dams altogether. My role was to ensure they were heard, and to help them to communicate their position to the government, soon to take office, using the laws binding the state. My advice also extended to balancing rights between investment protection under bilateral investment treaties and state duties under other treaty obligations. Effectively, my expertise was intended to provide policy recommendations to the new NLD government based on the direct and cumulative impacts of Salween hydropower development and international law, which did not support such a project.

I arrived in Yangon on 6 March 2016 after a long flight and, on arrival, had a meeting with my clients to hear their concerns in person. I spoke to various NGOs and met with representatives of the communities in Hpa'an, Karen State, to collect information concerning the dam projects. In particular, I wished to hear whether there had been any type of consultation. I also visited communities living on the banks of the river, including community representatives from the Hatgyi dam site, to learn directly about their position.

Travelling in remote Myanmar was a rare privilege. Unlike Thailand or other South Asian jurisdictions, there had been little Western presence in that country for decades and one could see it in the customs of the people, and the untouched environment. No jeans but *longyi*, a sheet of cloth widely worn in Myanmar. Women's faces covered in *thanaka*, a pale yellow paste made from pulverized tree bark, worn as protection against the sun and to keep mosquitoes away. Men chewed betel nut. The beauty of the landscape was breath-taking. Emerging from decades of military rule and isolation, Myanmar was just opening its doors to the world.

We held a meeting with representatives of communities living off the river, in situ. We sat around a large table, overlooking the Salween. Protected from the sun by an improvised outdoor shade sail made of white cloth, we talked for a few hours. I took notes. They spoke about the likely impact on their lives: the prospective depletion of fisheries, loss of agricultural lands (including islands) to flood and erosion, the degradation of forests, and the salinisation of freshwater areas. With his hand pointing at the likely areas that would be flooded, a fisherman looked towards the river. The community belonged to low-income, uneducated ethnic minorities who relied on the river for livelihood and sustenance. They understood clearly the devastating impact such a project would have on the river and their livelihoods.

Most villagers living within the Salween watershed engage in agriculture and fishing as the prime sources of livelihood. Due to changes in water level, nutrient content, sediment load and salinity, large swaths of arable land would become unsuitable for crop cultivation. The irreversible alteration of river, delta and offshore ecosystems would threaten wildlife and vegetation in areas of remarkable biodiversity that span nearly the length of the country.

We also learned that, according to seismologists and engineers, the construction of dams and resulting reservoirs would have disruptive impacts on major fault lines and increase seismic risks, including the occurrence of earthquakes. In the event of an earthquake, a single dam collapse upstream would create a domino effect due to the cascade of dams planned downstream on the Salween and its tributaries. The result would be massive and uncontrollable flooding in eastern Myanmar, from Shan to Mon States. The advice of earthquake experts was that the building of dams should definitely not go ahead.

None of the joint-venture hydropower projects included provisions for revenue sharing with local ethnic communities, who would bear the negative impacts of these large-scale infrastructure projects. Nor were any provisions in place to compensate villagers for loss of land, crops and livelihood. Those we spoke to told us that they did not believe in the process of granting the Environmental Compliance Certificate for the dams, as there had been no transparency and no proper consultation with local people. They were adamant: local people would gain no benefit from the dams.

We took pictures with the local communities next to the river. I remember the scorching sun, the dust on the way back and the smiles of the locals, some living off the earth in ways that humbled me. On our way, we encountered women devoted to the task of collecting earthworms from the mud.

These were just a few of the affected communities, but the scale of the proposed developments was staggering. The planned Mong Ton dam, which would be the largest hydropower project in southeast Asia, if it were to go ahead, would threaten the lives, homes and property of countless communities in the Shan, Karenni, Karen and Mon States.

One of the messages I received was that the Salween dam projects were also fuelling tension and conflict between different communities. The people argued that the dams were not just going to create misunderstanding between ethnic peoples and the new government, but would also impact on ethnic armed groups and the peace process. As they pointed out, some of the dams would be constructed in an area where several ethnic armed groups operated. Government troops were using the pretext of providing security for the dams to expand their presence in these areas.

Having heard from the people who would be most affected, it was time to speak to those who had the power to make decisions. Back in Yangon we met a delegation from the NLD, who were to take over government in Myanmar, including the NLD member working on hydropower policy. We met in some private rooms in a restaurant that had been booked for our meeting. There was a solemn air when we arrived and shook hands with a number of officials who were in attendance. Sitting around a table, we discussed the legal framework surrounding these energy projects and advised on the legality of the previous government's actions. We were able to put forward the communities' position, which was that the new government should stop all plans to build dams on the Salween River, as the impact would negatively affect the lives of countless communities in Myanmar.

In the long history of opposition by the Burmese communities to the plans concerning the Salween, this meeting was a rare opportunity to convey the Burmese communities' position, informed with the language of international law, at the highest level. Sitting at the same table, we flagged the potential breaches Myanmar would incur and the devastating effects, if the projects were to go ahead in the manner envisaged.

But we know, of course, what has happened in Myanmar since. We now know of the fall from grace of Aung San Suu Kyi, who assumed the post of state counsellor upon the NLD coming into office, after she defended the Myanmar government against allegations of the Rohingya genocide. In early 2021, the NLD were overthrown when a *coup d'état* put a military junta back in power. The Salween plans, with all their devastating effects, have been ignored by the rest of the world. The plans were stalled, but they remain on the table.[21] This is in stark contrast to the direction of travel concerning decommissioning of dams worldwide, as the Klamath river story reflects. But my experience has taught me that the soft power of people coming together and defending rivers has the same immense power as water. My mother, who grew up observing the ways of nature in the Andes, taught me this principle. 'Never forget that water pierces the stones,' she told me. Communities' persistence and soft determination, just like the water, have saved irreplaceable worlds.

The indigenous peoples in Myanmar continue working to keep the Salween River free-flowing.[22] They gather together along the river, and elderly Karen spiritual leaders perform a ritual to pay respect to the spirits that protect the river and forest every 14 March, on Action Day for Rivers. In the Karen belief system, humans are not the ultimate owners of the land and water.[23] The sun, the moon, mountains and rivers all have spirit owners known in the S'gaw Karen language as *K'Sah*.[24] To them, a river is alive.

These communities know that river systems are the zone of the Earth's highest biological diversity.[25] So far they have successfully opposed any dam building in the Salween, now empowered, as they are, with the language of the law.

Do rivers have rights?

The case of the Salween River shows that we need a shift in consciousness. You see, it is not just the Salween, or the rivers in the land of the Wayúus (as seen in the previous chapter), or the rivers in the ancestral land of the Ch'orti. It is everywhere. This ubiquitous degradation and pollution of rivers requires a new approach. I discuss this paradigm-shift in some depth in Chapter Nine, but here I want to introduce some initial thoughts.

Why do I place an emphasis on rivers? Rivers and other freshwater ecosystems are considered by naturalists to be 'the most vulnerable of Earth's habitats'.[26] New mechanisms for their protection are needed. This is starting to happen, but not quickly enough. Increasingly, rivers (and other non-sentient beings) are seen as having some fundamental rights.

The two key instruments of environmental law (the Stockholm Declaration[27] and the Rio Declaration[28]) postulate a *human*-centric approach to the protection of the environment. Today, however, there is increasing recognition that forests, rivers, seas and others entities are legal interests in themselves. That means, absent of risk to people, a river can still have rights of its own. This approach is about protecting nature not just because of its utility for human beings, or for the effects that its degradation could cause on other people's rights (such as health, life or personal integrity), but for its importance to the *other* living organisms with whom the planet is shared, who are also deserving of protection. It is a way of thinking that argues that the extinction of ecosystems should be considered 'unacceptable by civilised peoples'.[29]

In recent years, the Inter-American Court of Human Rights[30] acknowledged this evolving tendency in contemporary law to recognise legal personality, and therefore rights, to Nature, not only in judicial cases but also in constitutional systems.[31]

Now, some may find such a development a marginal, quirky or even an odd concept. 'Features', as part of the land, are what one buys and sells. Objects. They can't possibly be 'persons' in a juridical sense. What sorts of things have rights? Do they have to be sentient or in some sense living? (Assuming, for the sake of argument that there is no life in a river.) One could possibly be inclined to think so. But surely that is not the case. In the history of humanity, slaves were not legal persons despite being sentient. They were human, but treated as objects: they were bought and sold. Legal history shows, likewise, that women had to fight to be treated as 'persons'. In *Bebb v Law Society*, a case from 1913, the Court of Appeal in England and Wales told Gwyneth Bebb, a woman (in the words of the judgment 'a spinster') who aspired to be a solicitor, that she could not, because she was not 'a person' within the meaning of the Solicitors Act 1843. It was mentioned, incidentally, that married women (and it was noted Bebb could marry) 'did not have an absolute liberty to enter into binding contracts', and could not therefore 'contract [eventually] with her clients'.[32]

When I read this judgment for the first time, I read it in disbelief. The reasoning perplexed me. For example: 'There has been [a] long, uniform and uninterrupted usage which is the foundation of the greater part of the common law of this country, and which we ought, beyond all doubt, to be very loth to depart from'; 'no women has ever been an attorney in law. No woman has ever applied to be, or attempted to be, an attorney in law.'[33] This reasoning, out of another century, triggers in me hysterical laughter. It was a world in which 'it was so' because 'it has always been so'. It was a world that was stuck in believing in immutable laws. Of course, this idea of immutable laws was wrong. Gwyneth Bebb did not win her case, but her challenge triggered a shift in consciousness and in the law. In December 1919, the Sex Disqualification (Removal) Act 1919, effectively removing the disqualification of women from the exercise of the law on account of their sex, received Royal Assent. In effect, women became, then, 'persons'. In December 1922, the first woman entered the legal profession as a solicitor in England and Wales. A month earlier, Helena Normanton became the first woman to practise as a barrister in England. Yet it was not until the Equal Franchise Act of 1928 that women over twenty-one were able to vote and so women finally achieved the same voting rights as men. It is not so long ago. Until recently, women did not have the right to have a bank account in England; it was, in fact, only in 1975 that a woman could open a bank account in her own name.

So what emerges is a progression in the acquisition of rights by different recipients of rights. The law appears fluid and not static. What anchors rights and what pivots a change? Values that society considers fundamental at a given time.

Let's trace some central notions. To understand the rationale of full 'personhood' underlying the Bebb case, I turn to something that piqued my attention. Full personhood appears, in Bebb's reasoning, to be tied to the capacity to buy and sell, to holding property in one's own right. Curiously, it appears that the ones that were endowed with rights, full rights, were only men, men that could own property. Should the notion of 'property' alone (or rather of being proprietor) have such a high-ranking status in a legal system? Life, surely, is a higher-ranking notion in any system of law. Clearly this was asserted as such after the Second World War, as I discuss fully in Chapter Nine. Why, then, should a river not be assigned rights to protect the life it carries? Why should human life be the only life worth protecting? Is it not a fact that human life can only subsist in any way in the context of the survival of *other* life? For example, human life is not possible without water. Water is alive.

Why do we perceive human beings as entities separate from Nature? Such perception contradicts even rational Western thought. When, in his seventeenth-century philosophical treatise, *Ethics*,[34] Benedict de Spinoza, admittedly 'one of the most important and famous philosophers of all time',[35] reflects on substance, he postulates that Nature is 'necessarily infinite' and that everything in existence is suffused with its essence. He argued that the entire universe was but 'a one physical embodied substance' and that all things in it were necessarily interconnected aspects of that single substance.[36] Spinoza identified a oneness in the universe, in existence. There is no distinction between the creator (Nature) and the creation (i.e. a river, human beings, etc.), he posited. Nature (or God, which for him are interchangeable notions) permeates it all. Albert Einstein, centuries later, would agree. He famously said: 'I believe in Spinoza's God who reveals himself in the orderly harmony of what exists.'[37] Spinoza's radical thinking was the forerunner of the Enlightenment. Yet we may still fail to understand the fundamental premises of the oneness of the natural world of which we are part.

Of course, one may be tempted to say that, within Nature, a right-holder can only be an entity that possesses reason. We may have retained such notions from Orthodox Aristotelianism that conceives 'a scale of beings in the universe'.[38] Human beings, possessing reason, would be at the top of the scale. By contrast, in Spinoza's thought, there is no such 'top of the scale' notion. He argued that 'it is the supreme law of nature that each thing strives to persist in its own state so far as it can'.[39] Human beings, deer, birds – all have the natural urge to persist. One could add, the *natural right* to life. Spinoza, therefore, recognised 'no difference between human beings and other individual things of nature, nor between those human beings who are endowed with reason and others who do not know true reason nor between fools or lunatics and the sane.'[40] In our modern societies, in effect, those who have lost reason do have substantive rights. But losing one's mental capacities does not affect the substantial aspect of rights, for example right to dignity, right to life.

Possessing or not possessing reason does not appear to be, therefore, the crucial distinction to determine who is a right-holder. Note that, in the development of our system of law, society has granted rights to abstract entities, such as corporations and now even to software. Corporations have even been acknowledged to have 'human rights'. In Europe, a corporation is a legal 'person' that can go all the way to Strasbourg to ensure its right to property, or that its right to fair trial is not infringed upon.

In a recent lecture I attended, Michael Black, KC, highlighted the fascinating fact that the Northern District Court of California had held that the Ooki DAO (a derivative trading platform) was a 'person' under the Commodity Exchange Act and thus can be held liable for violations of the law'.[41] He remarked that the next development in our systems 'may well be legislation that attributes legal personality to certain software entities'.[42] We are at the liminal threshold of a world in which software entities can be a person. Why should – in that context – a river (a fundamental ecosystem in our planet) be denied rights?

A devil's advocate may question – are there thresholds when we talk about rights of Nature? Does a single ant have rights, for instance, or is it the ecological system that the ant lives in that's important? I would say that ants as a species are to be protected, as much as the ecological system. They are not separate. One is immersed in the other. One lives in the other. Ants have a function in that 'oneness'. Ants turn and aerate the soil. They allow water and oxygen to reach plant roots. They shape biodiversity. Systems are interconnected. We may be unaware of all the interactions that may be taking place within said ecosystems. You remove a species and the system will invariably change. In North America, when they reintroduced the wolf, the grazing behaviour of deer changed and, as a consequence, rivers changed, a knock-on effect you might never have imagined.

And a single ant? I would suggest that if this question came to you, this is the type of perspective one has to move away from. The one that isolates. That is precisely what Alexander von Humboldt realised in South America. Up to that point science had been focused on identifying endless taxonomic units. Instead, he was interested in the connections. Individual phenomena were only important 'in their relation to the whole'.[43] I can point out to you, however, that in our own current system, not even the right to life of humans is an absolute right. But when a species is threatened with extinction, when we witness the disappearance of bees, of earthworms, and we know what they do – for instance, what earthworms do to the soil – the extinction and the decline of biodiversity ought to be fundamentally wrong, because we humans cannot create it.

What about the quality of sentience? Why do we assume that a river can't suffer? If it can die, surely, you can witness its suffering. The suffering of the salmon, of the species that form part of it and that die with it. But, as seen above, the lack of sentience is not the critical element to disqualify an entity as a right-holder. Corporations are not sentient. Yet, they have rights in our systems of laws.

In the midst of an overwhelming loss of biodiversity on Earth, there is currently a turn to new paradigms.

It is not surprising, therefore, that recent examples of the protection of nature in its own right by judicial means have arisen in

Ecuador,[44] Colombia,[45] India,[46] Canada[47] and Peru,[48] where rivers have been granted rights or legal personality. In Colombia, the Constitutional Court declared the Atrato, a river that flows through the globally recognised 'biodiversity hotspot' of Chocó, as an entity 'subject of rights' and ordered the government to clean its waters.[49] In New Zealand, a similar protection has been conferred to a river by law. The Te Awa Tupua (Whanganui River Claims Settlement) Bill conferred legal personality on the Whanganui River, granting it the right to be treated as a living entity.[50] The move would reflect a Maori tribe, the Whanganui iwi's, 'unique ancestral relationship with the river. Iwi who lived along the river not only relied on it as an essential food source, but held with it a deep spiritual connection.'[51]

In his book *Deep Rivers*,[52] the Peruvian writer José María Arguedas, who lived among indigenous Andean people and spoke Quechua, gave a glimpse into the Andean cosmovision, whereby 'all parts of nature are alive'. Rivers, rocks, and mountains (known as Apus[53] in Quechua) are all considered a living part of nature. It is a way of seeing the world that I have seen echoed in a documentary chronicling the fight of a community in Cajamarca, in Peru's high Andes, against Yanacocha, Latin American's largest goldmine. In this film, *Daughter of the Lake*, a woman talks to the Conga Lake – threatened by the gold-mining project – as if it were a living entity.[54] Mayas, and other indigenous groups like the Kogis, whose world-view I explore in Chapter Seven, hold similar views.

It is therefore not unanticipated that constitutions in Latin America reflect such principles. The preamble of the Political Constitution of the State of Bolivia, issued in 2009, establishes that:

> Since the beginning of time mountains rose up, rivers travelled their courses, lakes were formed. Our Amazon, our Chaco, our high plateau and our plains and valleys were covered in

> vegetation and flowers. We populated this sacred Mother Earth with different faces and from then on we understood the existing plurality of all things and our diversity as human beings and cultures.[55]

The same constitution anticipates that:

> People have the right to a healthy, protected and well-balanced environment. The exercise of this right must enable individuals and communities of present and future generations, as well as other living beings, to develop in a normal and permanent way.[56]

Thus, the Bolivian Constitution acknowledges an inextricable relationship between human rights and the natural world, and regards the Earth as 'sacred'.

For its part, the Constitution of the Republic of Ecuador acknowledges inherent Rights to Nature:

> Nature, or Pacha Mama, where life is reproduced and occurs, has the right to integral respect for its existence and for the maintenance and regeneration of its life cycles, structure, functions and evolutionary processes.
>
> All persons, communities, peoples and nations can call upon public authorities to enforce the rights of nature. To enforce and interpret these rights, the principles set forth in the Constitution shall be observed, as appropriate.
>
> The State shall give incentives to natural persons and legal entities and to communities to protect nature and to promote respect for all the elements comprising an ecosystem.[57]

These are the first legal recognitions of rights of Nature in modern legal systems. They follow an eco-centric approach, whereby Nature is not treated as a mere commodity but as a subject of rights. In his 1836 book *Nature*, Ralph Waldo Emerson devoted an entire chapter to the Western approach to Nature as a commodity. He questioned: 'Why should not we also enjoy an original relation to the Universe?'[58] He thought that we have lost this original way of seeing, of star-reverence, where 'intercourse with heaven and earth' becomes part of our 'daily food'.[59]

To an extent, a declaration of rights only takes place in a context of major violations of said rights. What is self-evident does not need to be declared. Ask indigenous peoples about 'Rights of Nature' and they would tell you that Nature simply *is*. And it is respected and revered. But our inability to cohabit harmoniously within 'the community of life on Earth' – to borrow a phrase from *Wild Law*'s author, Cormac Cullinan[60] – has created human-induced changes to the environment that are so destructive that they are challenging the viability of life on Earth. As a consequence, our modern world is turning, now, to the enunciation of Nature's rights, just as the way, after the Second World War, the world turned to enunciating the self-evident quality of human rights, in an urgent attempt to protect them. It is a way of acknowledging the interconnectedness of the web of life of which human life is only part, and accept the inherent value of all living organisms and of biological diversity itself.

You may wonder, when Nature is granted rights, who polices those rights? So far there have been two main models developing. In most constitutional examples, anyone has the right to seek the protection of the Rights of Nature, including having a procedural standing in courts on behalf of Nature. I discuss that example in the Los Cedros case further on in this book. In the ad hoc model (cases in which rights of nature have been declared by courts, i.e. the Atrato River model), or by local governments (i.e. the example of the Magpie River in Canada), a guardianship system has been adopted. A landmark judgment by the Constitutional Court of Colombia, on 10 November 2016, in the Atrato case, declared that the Atrato River was a subject of rights that entail 'its protection, conservation, maintenance and in the specific case, restoration'.[61] The court provided that the Colombian State, together with the ethnic communities that live on the Atrato river bank in Chocó, were to exercise the guardianship and legal representation of the river.[62] The court ordered that this system of guardianship had to become operational within three months of the notification of the judgment.[63] The court set that the commission of guardians would count upon an 'advisory panel' or (expert panel/*equipo asesor*) comprised of two organisations with experience in the protection of rivers, and some entities (universities, research centres and other institutions with relevant knowledge) to support the protection of the Atrato and its basin.[64] The court gave a term of one year, for the commission of guardians to design a plan to decontaminate the water sources of the Chocó – beginning with the Atrato River basin and its tributaries – the riverside territories, recover its ecosystems and avoid additional damage to the environment in the region'.[65]

While rights of rivers, or rights of Nature, may appear foreign to a Western approach, they are proving effective in the fight to protect the natural world from extractive industries, and exploitation, across the Latin American region and beyond. To go back to the Salween example, in a plural society, where indigenous communities like the Karen with their belief systems exist, shouldn't a constitution accommodate their visions of the world and acknowledge the sacredness of its rivers? These are indigenous peoples, who enjoy a particular relationship with the land (including rivers), recognised by international law. The current Myanmar Constitution does not contain rights of Nature. But should it? Would that have not avoided the predicament of the Salween River today?

In 2020, a soft law instrument, called the 'Universal Declaration of the Rights of Rivers', was developed by the Earth Law Center, an NGO. It acknowledges that rivers perform essential ecological functions, play a vital role in the Earth's hydrological cycle and 'that the viability of rivers to play this role depends on numerous factors, including the maintenance of surrounding river catchments, floodplains, and wetlands'.[66] The Declaration draws from victories for the rights of rivers worldwide as well as scientific understandings of healthy river systems.[67] The preamble expresses concern 'that humans have caused widescale physical changes to rivers through dams and other infrastructure, which includes the construction of over 57,000 large dams worldwide that impact over two-thirds of all rivers, resulting in fragmented habitats, reduced biodiversity, imperilled fish populations, exacerbated climate change, and retained sediment and nutrients that are fundamental to downstream ecosystem health'.[68] The Declaration recognises that all rivers have 'the right to flow'.

The means used to protect rivers have now spread to the protection of forests. A recent example is the Supreme Court of Colombia's granting of legal personality to the Colombian Amazon region, a rainforest 'roughly the size of Germany and England combined',[69] which has seen deforestation rates increase by 44 per cent from 2015 to 2016.[70] The case, brought by twenty-five children and young people with DeJusticia (a Colombian NGO), effectively obtained that urgent measures be taken by local government and central government to protect the Colombian Amazon from deforestation.

Do we have the right to Nature?

The concept may not prove important only for forests. Think of the value of trees in a city. In recent years I have seen communities fight for the lives of trees in their cities. I learned, some years after my work with the Salween River, of a community in Southend-on-Sea, in Essex, England, fighting for the life of a single tree. Chester, a plane tree, was threatened with being felled, because it stood in the way of 'development'. The tree was looked upon as expendable, and not as a living entity.

I compared the tree canopy cover of Southend-on-Sea and that of The Hague, a city I visit often because of my work, using Google Maps. The difference was striking. The Essex town was tree-deprived. Chester was the only tree in an area of pure concrete. By comparison The Hague was lush and green. In fact, Southend Council itself had acknowledged that Southend's canopy cover was below the national average in the UK. In essence the population in Southend were adversely affected, discriminated against, I would say, by the lack of trees. To me they were entitled to trees and, by threatening to fell one of the last mature trees in their town (without considering alternative ways of preserving it), the community was indirectly discriminated against by comparison with the standard of living of the average urban area in England.

But it is not just the service we obtain from a tree. It is about the essential unity of all things. Our humanness is achieved via the interaction with our natural surroundings. 'I understand what you are feeling,' I told a member of the Southend community over the phone, as they explained to me why they valued Chester. A tree next to my house fell during a storm on 2 January 2024 (just a step short from my door) at dusk. It was a shock, and strange not to see, next morning, its bark and canopy from my window. I used to feel as if I were on top of that tree sometimes, given its proximity. It was a 150-year-old oak. The tree must have been a remnant of ancient woodlands in this bit of Surrey (to where I had moved) and I chose my new home because of the trees surrounding it. So, my beginning of 2024 was coming to terms with its absence. Do you know the term *solastalgia*? It was coined by Glenn Albrecht, a philosopher. According to him, solastalgia is when your endemic sense of place is being violated. Is this what the communities of the Salween and the community in Essex were also trying to avoid?

My sense of *home* had irreversibly changed with the demise of my tree. The tree was important to me. I had developed a connection. I used to see it daily from my window and I greeted 'them' daily (my partner explained to me that my oak tree was a 'she' and a 'he' at the same time). It was covered in moss and lichen (as are most trees around here), which the sun would light up wonderfully. It was really beautiful. Lots of little creatures – blue tits, tree creepers and other birds – use to hang around the tree. It was the first thing I used to see in the morning. And all of a sudden it was gone. I was in shock. Watching it lying in the road, blocking it, the next morning, felt as if the moon had suddenly fallen. It was like watching something formidable dying. Its fallen branches reached the entry of my house. I appreciated every minute of its company. I dreamt that week that the tree was still there under the moonlight and its mossy branches entered through the window. It was so real. It was the last time I felt its presence. My tree's disappearance filled me with a sense of loss.

I kept a piece of it, a small mossy branch, a bitter-sweet memento, in my garden. I felt like I'd lost a friend. Sadly, it was planted in the garden of my neighbour, so nothing will replace it. He said that it used to block the view of the road for him. So, his sense of place has also changed with the loss of the tree. He himself maybe did not know how important the tree was for his own sense of happiness.

There may be a sea of differences between the Karen people in Myanmar fighting for their river, and the community in Southend fighting for a single tree, but ultimately I see a similar connection to a living entity that means a lot to them and that they wish to protect. I see a strong sense of kinship.

THREE

The Dark Business of Light in the Land of the Birds

'Rivers, for us, represent the serpent, life, what is female, what is male, continuity. When a river is blocked, the flow is cut, the connection to the land is cut. The water is to the earth what blood is for us.' This was what I was told by Omar Jerónimo, a Ch'orti' leader, as he attempted to explain their perspective and opposition to hydropower projects in the Jupilingo River, which is located in their ancestral territory.

In 2015 I arrived in Chiquimula, Guatemala, a place somewhere near the border with Honduras. Chiquimula, a word of Aztec origin, means 'Land of Birds'. Its forests house a large number of stationary birds and also has areas sought after by migratory birds. You can find kestrels, chorchas (orioles) and azulejos, motmot, oropéndolas, woodpeckers, and many more.[1] You can also find peculiar species of spiders, such as the *araña tejedora espinosa* (who, against the sky, suspended in their cobwebs, look like tiny aerospace ships with orange spikes), and curious insects like the *chinche linterna negra*, which has long waxy tail filaments.

The Ch'orti' case is especially important to me because of what it has taught me about issues surrounding land, collective rights and consent. It also posed fundamental questions about what 'development' is, and the mega schemes concerning energy generation (not for living needs but for profit) that threatened entire areas with destruction (while not serving the communities living in that area). It is as if, for such grand schemes, the presence of people in large parts of the world were inconvenient. In the way. In the way of what? Of 'progress'? Progress for whom?

Wait. But what is *progress*?

—

I had arrived in Chiquimula as part of an independent fact-finding mission, together with a former UN officer who had worked for many years on indigenous rights and another barrister with experience in corporate responsibility. We had been invited by the Central Campesina Ch'orti' Nuevo Día,[2] an indigenous association that works alongside indigenous Ch'orti' communities in Jocotán, Olopa and Camotán, facing threats and violations to land, environmental and cultural rights by hydroelectric and mining projects in their territories. The Ch'orti' were in conflict with two companies owned by America Trans Group (Las Tres Niñas and Janbo), which planned to build hydroelectric dams on the Jupilingo River. As the Ch'orti' we interviewed explained, 'in fifteen kilometres of river, they want to build three dams'.

The purpose of the mission was to investigate and record observations and findings regarding the conflict and make recommendations in a written report. After many months of preparation, reading background information and Skype sessions to secure meetings with different local and central governmental authorities, we arrived in Guatemala. From 2 to 9 May 2015 we met with the relevant people involved in the conflict (including representatives of the Ministry of Energy and Mines, which is ultimately responsible for giving permission for the building of hydroelectric dams, mines or other large energy projects in Guatemala) and other organisations able to provide information for understanding underlying systemic issues.

This was my first fact-finding mission and visit to Guatemala, a country I had read a lot about. The streets of Guatemala City, the capital, leading to Plaza de la Constitución, were full of demonstrators. As our fact-finding mission to Guatemala was taking place, the country was being shaken by public protests against corruption at the highest level. On 9 May 2015, President Otto Pérez Molina accepted the resignation of the country's vice-president, Roxanna

Baldetti, following allegations of corruption, and the ministers for environment, interior and energy, as well as the heads of the intelligence services and the central bank, were arrested on corruption charges. President Pérez Molina was also called upon to resign, accused of corruption as well as of committing crimes against humanity during the civil war. The political crisis taking place during our visit, together with the multitudinous campaigns for the presidential, legislative and municipal elections due to take place in September 2015, only served to underline the ongoing challenges facing Guatemala.

Guatemala had endured thirty-four years of civil war, beginning in 1962. Under the auspices of the United Nations, peace was negotiated between the government and the Unidad Revolucionaria Nacional Guatemalteca (URNG) in 1996. The Commission for Historical Clarification set up to investigate and take testimonies from victims of human rights violations concluded that some 200,000 people were murdered or forcibly disappeared, mostly indigenous peoples. It is estimated that 83 per cent of the total number of victims during the armed conflict were Mayans.[3] More than 90 per cent of the crimes committed during the war were committed by the state or paramilitary groups under its control. The commission concluded that acts of genocide were committed against the Maya Ixil people in the period from 1980 to 1983 when General Rios Montt was president and the then president, Pérez Molina, was Director of Military Intelligence.[4]

At some point during my stay in Guatemala City, I went to the museum of history. I was surprised. Nowhere among the displays of the museum was there any acknowledgement that a civil war had taken place in Guatemala or that acts of genocide had been committed against Mayan indigenous peoples. It was one of the most shocking demonstrations of denial I have ever encountered.

By 2015, the underlying causes that led to the conflict remained largely unaddressed. Despite some reforms and efforts to address matters such as impunity, judicial independence, fairer land distribution and poverty, the country remained largely under the control of an unaccountable military and private sector, with little of the Peace Accord implemented in practice.[5]

The Ch'orti'

The Ch'orti' are one of the indigenous Mayan peoples. They live in south-eastern Guatemala, north-western Honduras and northern El Salvador. They number approximately 51,000. The majority of Ch'orti' live in Guatemala, including large numbers in the region of Chiquimula.[6] They speak Maya Ch'orti'. I confess that, prior to this mission, I did not know the Ch'orti' existed.

In common with other Mayan peoples, the Ch'orti' have a special relationship with the Earth. I must stress, though, that when you ask indigenous peoples about it, they never refer to it as 'nature', they always refer to it as the territory (*el territorio*), even if they consider that the territory has a life of its own. When asked what they mean by their territory, they explain that it is the land where they were born, where their ancestors were born and where their children will be born: 'It is where we originate, where we all live, where we grow up, where we eat, and sit to rest.' They told us: 'The land is, and has always been, ours. It is the heritage left by our ancestors.'[7] Their understanding of *territory* comprises the earth, the rocks, the forest and the river – all of which require their care.

The river at the heart of the dispute was the Río Grande de Jocotán, or Jupilingo River (as it is known by the locals), which traverses Ch'orti' ancestral lands in Honduras as well as the Chiquimula region. Because Chiquimula belongs to the region known as the *corredor seco* (or dry corridor), where there is little water or vegetation, the Ch'orti' describe the Jupilingo as a miracle: 'What is the Jupilingo River? It's a miracle to see the river because for the last three years we have lost our harvest from drought. It's a miracle. It is our only natural resource, our source of life. Our lives depend on it.'[8] We were told of their dependence on the Jupilingo: 'When there is no water, we rely on the river.'[9] The Ch'orti' people also described having very close links to the forest. They said all Ch'orti' people have a forest next to them – not to exploit, but to live with.

They support themselves mainly through agriculture on the mountain slopes. As they put it: 'We eat from the land.'[10] They plant different varieties of corn, beans and pumpkin in the *milpa* system, an ancient and sustainable way of farming. Their crops are vulnerable to the frequent droughts in the region. The river is essential in supporting these crops as well as providing water for cattle.[11] It also serves the Ch'orti' in other essential ways. Families collect water for cooking from the river, they bathe in the river and women bring their clothes to wash in it.

The first morning, after our arrival in Guate (as the capital is commonly known), I came down to have breakfast and met Claire, the other barrister joining the mission. She was tall, vegetarian and easy to talk to. Her mother tongue was French, and so she was able to understand Spanish and speak it to a level at which she could communicate. We ordered avocado and corn, probably the best avocado I had eaten in years.

I was happy. I was in Latin America again after some years of absence. It was pleasing to hear Spanish being spoken in the streets, to see a familiar chaos around me, perceive a sense of spontaneity in the air. Claire had been involved in large tort cases (which involve harm arising from non-contractual scenarios) in the English courts, relating to environmental issues abroad, an area of law I had also developed in my practice. We did not know what to expect from our visit to the territory of the Ch'orti'.

According to the 2011 National Census of Guatemala, indigenous peoples number approximately 40 per cent of the total population in a country of 14.4 million.[12] Other sources estimate that the indigenous peoples constitute more than 50 per cent of the national population. According to the Inter-American Commission on Human Rights, 'the majority of the Indigenous population in Guatemala is Maya'.[13] There are twenty-one distinct indigenous peoples of Mayan descent.

There has been a history of discrimination against the indigenous populations in Guatemala since colonial times, as acknowledged by inter-governmental organs (both at UN level and at regional level in the Americas).[14] The Commission noted:

> The present condition of the Indigenous populations in Guatemala is the result of the long colonial oppression process against the Mayan people as of the sixteenth century, consolidated under the liberal national Government during the nineteenth century, upon the constitution of a governing class that based its power on large rural land property and the exploitation of Indigenous labour, within the framework of authoritarian and patrimonial regimes.[15]

The Inter-American Commission on Human Rights (an organ of the Organisation of American States), in particular, has pointed out that discrimination against indigenous populations is 'undeniable' in Guatemala.'[16] As in other Latin American countries, the indigenous population is notably poorer than the non-indigenous population[17] and life expectancy for indigenous peoples is thirteen years less. They have less access to education and are disproportionately underrepresented in secondary and tertiary education. According to one study, only 14 per cent of indigenous girls attend primary school.[18] Further intergovernmental reports note that 70 per cent of indigenous children in Guatemala are malnourished.[19]

In addition to the generally disadvantaged situation of indigenous peoples in the country, there are major issues of human rights arising from large-scale projects, particularly in relation to resource extraction on indigenous peoples' lands. As noted by the UN Special Rapporteur on the rights of indigenous peoples:

> the business activities under way in the traditional territories of the indigenous peoples of Guatemala have generated a highly unstable atmosphere of social conflict, a situation recognised not only by the affected peoples but also by the public authorities, civil society and the companies themselves. It seems that this situation has not only had harmful repercussions on the indigenous peoples and communities but has also made it difficult for the Government and for business people themselves to promote investment and economic development in Guatemala.[20]

The legislation passed by Congress to facilitate investment in large-scale projects has not included any recognition of the particular situation of indigenous peoples, their interests with regard to their traditional lands or their right to be consulted. This is the case with the Mining Act, the Hydrocarbons Act, the Forestry Act and the Electricity Act.

In general, there is neither a recognition of indigenous peoples' rights over their traditional lands nor any procedure in place to ensure that good faith consultations (a notion discussed in Chapter One) are undertaken. Furthermore, when indigenous peoples have held their own consultations, they have not been recognised. None of the more than seventy community-based consultations carried out by communities affected by large-scale projects have been given consideration by the government, and the Constitutional Court has declared them non-binding.[21] These community-based consultations have been initiated because of the failure of the government to respect the engagements required under ILO Convention 169 on indigenous and tribal peoples, an international treaty under the auspices of the International Labour Organization that deals exclusively with the rights of these peoples.[22]

Hydroelectric dams in Guatemala

The Chixoy hydroelectric dam was built in 1985 and is the largest in the country. It caused the forcible displacement of about 3,500 persons from thirty-three Maya Achi communities and resulted in the massacre of 440 people. In the absence of national action to recognise the human rights violations caused, or provide compensation for the loss of homes and livelihoods, the communities eventually sought a decision from the Inter-American Commission on Human Rights, which recommended that the Government pay compensation. In 2014, the then president rejected the compensation plan, stating that the country was not bound by rulings of international bodies. The president, following pressure from the US, the World Bank and Inter-American Development Bank, which had partly funded the dam, only finally agreed to a compensation package in October 2014.

Thirty years on, hydroelectricity projects are still a major source of conflict. According to a 2015 report by the UN High

Commissioner for Human Rights on activities of his office in Guatemala, communities opposed fourteen of the thirty-six hydroelectric plants being constructed or planned. The report continues: 'In view of the State's lack of intervention to guarantee that indigenous people are fully informed and consulted, some companies made direct contact with the communities, which in many cases led to divisions within communities, given the failure to observe their traditional forms of organisation and decision-making.'[23]

The Ch'orti' communities, in particular, were concerned about two hydroelectric projects: one in El Orégano and one in El Cajón del Río. These projects were owned and planned by two different companies, Las Tres Niñas SA and Jombo SA. We met with Mr Kenneth Jangezoon, who maintained that he was the sole owner of these companies. Mr Jangezoon was a tall man with light brown eyes. We appreciated the fact that he made time to meet us. We obviously wanted to hear his position.

Las Tres Niñas SA planned to build the El Orégano hydroelectric dam on Río Grande (Jupilingo River) between Zacapa and Chiquimula. The application for the licence from the Ministry of Energy and Mines specified a height for the dam of 120 metres[24] and a production of thirty megawatts. However, Mr Jangezoon told us that the project had been downsized six months prior to our meeting in May 2015 to a dam of forty-nine metres. These measurements made a difference for regulatory purposes, as we learned. Jombo SA, for its part, planned to build a dam in Cajón del Río with a height of fifteen metres and a production of four megawatts. This dam did not appear to be subject to any regulation, because dams with a production of less than five megawatts do not require a licence in Guatemala, and can be built by anyone, as we were informed by officials from the Ministry of Energy and Mines.

Contrasting with what Mr Jagezoon had told us, Nuevo Día

claimed that both companies are owned by a larger parent group, called Trans America Group.[25] We were able to later verify that the companies do feature on the website of Trans America Group. This company, it is claimed, is owned by the Gutierrez-Bosch family, one of the richest in Central America.

Territories

'Land conflicts are one of the many problems that especially affect Indigenous populations in Guatemala', noted the Inter-American Commission on Human Rights in its latest report concerning Guatemala.[26] This, it continued, is derived, among other things, from a lack of actual recognition of indigenous territory and lack of an effective registration system that recognises ancestral territory enabling the protection of land belonging to the indigenous populations.[27] When it comes to the right to property, as it was put by a representative of the civil society: 'The law is for whites, not for indigenous people.' The denial of ensuring the right to property of indigenous peoples, however, from a legal perspective amounts to discriminatory treatment.

Although the Maya Ch'orti' have historically occupied the land they inhabit as indigenous people, and as such have a collective right to their land and territory, this right primarily derives from customary law, from the fact that they have occupied those lands since ancestral times, rather than from the actual registration of their land as the collective property of the Maya Ch'orti' *qua* indigenous people.

Land registration is tied to the prior legal recognition of the indigenous community. Gaining legal recognition (i.e. legal personality) first is therefore a necessity. Yet in their own municipalities it has proved to be an uphill battle for the different Maya Ch'orti' communities. Second, the land rights of indigenous peoples in Guatemala continue to be circumscribed by legislation below constitutional level, which fails to implement the rights to which indigenous people are entitled under constitutional and international human rights law.[28] In particular, the Guatemalan civil code fails to regulate indigenous communal property.

The 'invisibility' of indigenous peoples

'Yes, the Ch'orti' do exist'[29]
A Ch'orti' villager

I was puzzled by the overwhelming reality of the existence of the Maya Ch'orti' people as a distinct collective and yet their 'invisibility'. For the purposes of the law, they did not exist.

'Invisibility' is the topic of an ineffable story, *Garabombo the Invisible*,[30] by Peruvian author Manuel Scorza. The eponymous Garabombo spends seven days sitting down in the Subprefectura, to record a complaint against the big landowner (the *hacendado*). The authorities come and go, but they don't 'see him'. At the beginning he assumed they were just busy, but as days went by, he became aware of the phenomenon of his 'invisibility'. 'But I see you!' a member of his community told him. 'You are of our blood, but the whites don't see me,' he answered. Scorza had depicted the invisibility of the indigenous in Peru of the 1960s, to show the absolute estrangement between these two worlds, white and indigenous. I admit, Scorza's work had an impact upon me and fuelled my interest in becoming a lawyer. I wanted to make those like Garabombo visible; I wanted the law to respond to them.

Back to the Ch'orti' – from a legal point of view, their existence as indigenous Maya Ch'orti' people has been a *de facto* and not a *de jure* (in practice rather than from the law) reality. Because of this lack of juridical recognition, they could not act procedurally as subjects of rights and obligations; they could not take legal or administrative action. The Constitutional Court of Guatemala has set criteria that 'juridical persons must accredit their existence; their de facto existence in itself does not give it legitimacy'.[31]

So, in 2014, seven Maya Ch'orti' communities began the legal process of formal recognition as indigenous Maya Ch'orti' in their municipality.[32] The community of Las Flores won formal recognition on 5 August 2014,[33] but at the time of our fact-finding mission, the applications for the other communities were still awaiting a decision and facing opposition from businesses and other parties.

Legislation that is supposed to enable rights under the Guatemalan constitution fails to recognise collective rights to land. As noted by the United Nations High Commissioner for Human Rights on the activities of her office in Guatemala, in February 2013, the Constitutional Court of Guatemala resolved a claim that challenged the constitutionality of part of the Civil Code because it did not regulate indigenous communal property.[34] Already in an earlier report, the High Commissioner had pointed out legislation that recognises 'traditional forms of collective land tenure, possession and use by indigenous communities' needed to be adopted.[35] In its decision, the Constitutional Court recognised that this was an 'issue that was pending in the legislative history of the country'.[36]

The Ch'orti' people followed a path many indigenous peoples that were stripped of their ancestral lands in Guatemala have followed, so as to ensure that their rights to property would be respected: they bought the land they are living on in 1878 (an injustice in itself), but this right has never been acknowledged by the authorities.

Some progress towards 'recognition of collective land management' in protected areas had nevertheless been made.[37] For example, six communities of the Sierra de Santa Cruz registered their collective property in the Property Registry Office (Registro de la Propiedad Inmueble).[38] So, despite the inadequacies of secondary legislation, the State of Guatemala could and is bound to use any legal and administrative means to respect its own constitution and international agreements made part of its constitutional corpus.

Collective rights

> Guatemala is formed by diverse ethnic groups among which are found the indigenous groups of Mayan descent. The State recognises, respects, and promotes their forms of life, customs, traditions, forms of social organisation, the use of the indigenous attire by men and women, [and their] languages and dialects.
>
> Constitution of Guatemala 1986, Article 66 (Protection of Ethnic Groups); Third Section (Indigenous Communities)

With the disappearance of authoritarian governments in Latin America, new constitutions have been adopted in many countries, recognising the plurinational and multicultural character of the state. From a situation in which the indigenous peoples of the countries were deemed merely citizens of the state with no distinctive juridical personality, there has been widespread recognition of indigenous peoples as collectives with accompanying rights, including rights to their traditional lands. This is the case, for example, in Bolivia, Brazil, Colombia, Ecuador, Mexico, Peru and Venezuela. In these countries indigenous peoples are recognised as peoples with distinctive cultures, languages and identities, rights over land and natural resources, as well as the right to consultation prior to the exploitation of non-renewable resources located on their lands.[39]

The Guatemalan Constitution of 1985 recognises ethnic groups[40] and although it recognises 'indigenous *communities*', it does not refer to them as indigenous *peoples*. The constitution requires the State to pass enabling legislation to provide land to indigenous communities that may need them for development[41] and to regulate the section of the constitution relating to indigenous communities.[42] This never happened. Nevertheless, the State has positive duties to respect the rights of indigenous communities and the duty *to promote* their way of life, customs, traditions and their organisations, languages and dialects.[43] The Guatemalan Constitution acknowledges that indigenous communities collectively own lands that historically have belonged to them and places a duty *to protect* such property on the part of the State.[44]

As part of the Peace Accords, the Agreement on Identity and Rights of Indigenous Peoples was signed between the government and The Guatemalan National Revolutionary Unit resistance movement (UNRG) in 1995. The Agreement recognises the Mayan communities as indigenous peoples, as it does the Xinca and Garifuna peoples. It also acknowledges the multi-ethnic, multicultural and multilingual nature of the nation. The Agreement calls upon the government to make the necessary changes to the constitution, which it has not yet done. It also calls on the government to establish mandatory mechanisms for consultation and, more specifically, requires the government to adopt measures to secure 'the approval of the indigenous communities prior to the implementation of any project for the exploitation of natural resources which might affect the subsistence and way of life of the communities'.[45]

If you are a European reader, it must be novel to hear about 'collective rights'. The question of collective rights was discussed at length in international human rights forums from the 1980s. Initially a number of states considered that human rights applied only to individuals and could not be extended to groups, or even that to provide collective rights to one group would diminish the rights of others. The adoption of the Declaration on the Rights of Indigenous Peoples in 2007, however, with the support of all member states of the United Nations, set aside this debate. The rights of indigenous peoples are considered to be of a collective nature and the rights elaborated in the Declaration are deemed essential for the continued development of their distinct cultures. Article 3 of the Declaration recognises the right of indigenous peoples to self-determination and further articles of the Declaration elaborate on how that right should be understood. The term 'peoples' refers to distinctive identities, histories, cultures, languages, ways of life, political and social organisations, beliefs, sciences and laws that are to be found among indigenous peoples.

But in the Guatemala I visited, there was a marked absence of recognition of indigenous identity in the discussions we held with governmental officials or representatives of the private sector. I was constantly reminded of the 'Garabombo the Invisible' story. This is a curious paradox, since the Mayan peoples have a long and well-recorded history, clearly pre-dating the Spanish and subsequent colonists, and the descendants of that civilisation continue to have attachments to their ancestral lands, cosmologies, sciences and legal systems. All remain a part of daily life. The failure to recognise in practice the collective rights of indigenous peoples, and in this case of the Ch'orti' peoples and their institutions, for collective decision-making by outside interest leads to the assumption that *individuals* from the community *can* enter into agreements on behalf of the group as a whole (as the companies seemed to assume). The companies, allegedly, were trying to get individuals to agree to sell their land. It resisted treating the Ch'orti' as a collective.

Rights to land and territory

Guatemala has one of the most unequal land distribution patterns in Latin America. The largest 2.5 per cent of farms occupy nearly two-thirds of agricultural land, while 97.5 per cent of farms occupy one-sixth of land. Despite the urgent need for clarity around the issue of land tenure within its domestic system, there remain multiple unresolved land disputes, and ineffective mechanisms to resolve them.[46]

The Ch'orti' had the right to property to their ancestral lands irrespective of delimitation, demarcation or registration of their land, something I noted in our report, and it seems the Ch'orti' noticed with delight. In the case of indigenous peoples, their right to their land and territory[47] is closely tied 'to a particular form of life of being, seeing and acting in the world', whereby indigenous peoples have a close relationship with their traditional lands and the natural resources found in such lands, since they are 'their primary means of subsistence' and an 'integral element of their cosmic vision, religion, and therefore, of their cultural identity'.[48]

In the *Awas Tingni* case, the first case in which the Inter-American Court of Human Rights examined rights of indigenous peoples centrally, the court held that 'for indigenous communities, relations to the land are not merely a matter of possession and production but a material and spiritual element which they must fully enjoy, even to preserve their cultural legacy and transmit it to future generations.'[49]

This is precisely what the Ch'orti' were trying to do.

In the *Awas Tingni* case, the court decided that the government of Nicaragua was not within its rights to lease lands to a foreign company where the indigenous peoples affected were able to demonstrate their collective rights over the property through long-standing, spiritual and material relations with their territories.

Despite this legal framework, in the discussions with representatives of local and central government and the company building the dams affecting the Cho'rti' communities, there was no recognition of indigenous peoples' collective rights. 'For us that is new, calling themselves indigenous communities is new; it is for tourists,' we were told by the CEO of the companies, when he was asked about whether he was aware of the rights of the Ch'orti' as collective rights.

The most illustrative example of the absence of recognition of the collective identity and interests of the Ch'orti' people by governmental and private sector representatives was the case of the so-called COCODES – Consejos Comunitarios de Desarrollo. The COCODES are government-sponsored entities for participation at the municipal level and are within a structure that is integrated at the departmental, regional and national levels.

We met with representatives of the COCODES in the municipality of Camotán, together with the local representative of the company building the dam. The representative of the company explained that the project had involved broad consultation and would bring significant improvements to the community in the form of roads, tourist activities related to the artificial lake that would be created, as well as other small projects to a value of 100,000 Quetzales (approx. £8,380) per community decided on by the people. The company representative said that the consultation had been in the form of newspaper articles, posters and radio spots, and noted that, in the community to be directly affected, 'only five or six were against the dam'. The representative even went on to criticise the non-governmental organisation (NGO) Nuevo Día for 'fomenting discord in the community by providing incorrect information'. The representatives of the COCODES were in favour of the dam and solemnly spoke in much the same terms as the company regarding the benefits arising from it. They claimed that no negative impacts had been felt in the community and no one had been displaced.

The Cho'rti' communities did not recognise the position of these representatives before COCODES as representative of the Ch'orti's views. They argued that the representatives of the COCODES had effectively split the community, undermined traditional practices of taking decisions together in the interests of all, and sown discord and confrontation. The impression was that the driving force for the appointment of these particular individuals as COCODES was the company itself seeking to use the formal participatory structure as a means to gain support for the project. Our delegation was made aware by members of the civil society that: 'the pattern is repeated; the company creates its own indigenous authorities.'

Of note also was the language Mr Kenneth Jangezoon used, which denied the identity of the Ch'orti' as a people with distinctive collective rights. The term preferred by him to refer to the Ch'orti' was often farmer or peasant – '*campesino*'. In a lengthy explanation, he suggested that the identity of the Ch'orti' people was diminished due to migration and that their community work traditions had been undermined by cash being sent by relatives from the USA.

In Jocotán

However, the Ch'orti', in common with other indigenous peoples in Guatemala and in the wider region, govern themselves by means of an indigenous council. This is composed of elected community members. We arrived in Chiquimula and met with representatives of the indigenous council of seven communities from the municipality of Jocotán (Las Flores, Matasano, Guareluche, Escobillar, Pelillo Negro, Guahiquel, Ingenio Guaraquiche) and seven communities from the municipalities of Camotán (Cajón del Río, Palo Verde, Lelá Obraje, Lelá Chancó, Shupá, Rodeo, Pilincas).

We had to drive for many hours outside the town of Jocotán. The sun was intense and the landscape had a rare beauty. The climate was dry and the landscape wild, with its vivid colours, and untouched. The municipalities of Camotán and Jocotán have steep hills, of which 80 per cent is denuded of vegetation. The forests which used to cover these hills were cut down over a period of 300 years by the Spanish conquistadores as a way of preventing the Mayans from sheltering in the mountains and attacking them in the plains. I was to learn from the Ch'orti's that the slopes of their territory were, as a consequence of this, highly erodible. There was concern that a restricted flow of water in the Jupilingo would be insufficient to wash away the sediment that comes off the denuded mountain slopes. This would eventually cause the river bed to become sedimented and the river to dry up.

For many miles you could not see any trace of human intervention. Along the path, sometimes, you could see indigenous people walking along the ways in their bright clothing.

Around one corner, below us, we could see the Jupilingo. It was a special encounter. You could only truly understand why the river was so important if you were standing there, in the midst of this landscape, under a scorching sun.

In an open plaza hundreds of Ch'orti' women, some carrying their children, and men, who had gathered together, coming from many villages, were waiting for an open assembly with us.

'And now, international *señores* are arriving,' said a voice on a loudspeaker.

Blinded by the sun, I was able to distinguish nevertheless hundreds of Ch'orti' men with their hats, and women wearing clothing of bright colours. This was the first of several meetings, including one in Camotán, to listen to the position of the Ch'orti'. The biggest was in Jocotán. We had discussions with the representatives of the Ch'orti' in a room and then further discussions in the plaza.

We learned that the region had been hit by several decades of drought, leading to poor crop yields and famine. The Ch'orti' were fearful of not being able to meet their basic needs due to the precarity of their sources of subsistence. Against this backdrop, the prospect of changing the hydrological features of the region caused them real concern.

For the Ch'orti, the rain, rare as it is in the *corredor seco*, was understood to come largely from evaporation of the vegetation and river. Drought was not so much the lack of water but the misdistribution of rain over the year. There was a fear that concentrating the water in one basin, and restricting its flow elsewhere, might change the spread of humidity and rain in the region, thereby aggravating the risk of drought. In particular, there was a concern that the Sierra de Las Minas, a large forested mountain range north of Chiquimula, would dry up, as it received rain from evaporation of the Jupilingo River.

There were, at the time, seven applications for hydroelectric dams on the Jupilingo, one of which had been approved. This did not include any dams for which approval was not required, namely dams that would produce less than five megawatts of electricity. There was, therefore, a real concern that the river would be exploited far beyond its capacity, and thereby destroy what little natural resource the Ch'orti' had.

The Ch'orti' also raised the concern that the concentration of water through hydroelectric dams would reduce the water available for agriculture around the river. The employment generated by agriculture would be lost, and a basic means of sustenance weakened. This would likely lead to further migration away from the ancestral lands and into urban poverty. About 15 to 20 per cent of the community also carried out non-commercial, artisanal fishing in the river. This means of subsistence would also be lost, they argued. In addition to the principal concern over drought and subsistence, there were the more prosaic worries about access to the river for everyday activities. The Ch'orti' use the river for washing and recreational bathing. They were concerned about losing the river for these amenities. It is their land. 'We did not come here, we were born here,' the community repeatedly informed our delegation. There was a profound sense of connection with their ancestors, who had lived here for hundreds, if not thousands, of years. Added to this, there was a consciousness that there would be nowhere else to go should they have to leave. 'Wherever people go, there is no longer any land. They [the companies] are already the "owners".'

The communities were also suspicious of the information given to them by the dam company as to what the impacts of the project would be. The impacts were always described as positive. 'This is what we bring, they say. They don't tell us about the consequences, they just say it's good; they don't talk about the disasters.' No downsides or risks were ever discussed with them.[50] The failure to acknowledge potential problems arising from the construction of a dam made the communities distrustful of the present discourse regarding the dam.

Despite having yet to be constructed, the dam project has already had a very real impact on the community by causing unrest and division. The active resistance of the community to the project had also led to community members being stigmatised and, in their view, unfairly prosecuted for alleged criminal activity. The community felt that their members were being unfairly criminalised and persecuted only because they were seeking to have their voice heard. The divisions stemmed from the attempts by the company to gain community acceptance through financial or other incentives. Those community members whose land was purchased by the company, or who had received some aid, or, allegedly, bribes, were supportive of the project. It appeared that, although they were in a minority, this had been sufficient to tear the social fabric of the communities in a way that was felt acutely and was difficult to repair. The minority was accused with having been bought off. They were seen as falling foul of the saying '*pan para hoy, hambre para mañana*' (bread today, hunger tomorrow).

The minority (the COCODES) argued that the community was resisting progress and development, and that it was misinformed. This had caused a climate of fear and suspicion. Outsiders could not enter the community without being noticed. The community was on constant alert for any sign that company officials had come to bribe or persuade the community through some other means, or that construction was about to start. Outsiders to the community who had not organised their visit with the community members were often stopped and asked what they were doing. The climate of suspicion, the divisions within the community, the clash with local government and the criminalisation of the community all had their roots in the failure to implement the rights of indigenous peoples recognised in the law of the country, in particular through its ratification of ILO Convention 169. Proper consultation of the community, and a recognition of its right to dispose of its territory and natural resources, would have led to an entirely different outcome. The dynamics of the relationship between the community, the company and local government would have looked very different if the existing legal framework in Guatemala were in fact implemented at national and local levels.

After the meetings I was exhausted. It may have been the heat; it may have been the copious amounts of information I had to absorb so quickly. As I went to sleep, I had the image of the Jupilingo running through verdant hills in my head.

The business model

'There is a business culture that is stuck in another century'
Civil society observation

Clearly, from our visit in Jocotán, we could not see any benefit arising from the project for the Ch'orti', and the companies had no understanding of the requirements of benefit sharing under the law. We met again with the companies' CEO, Mr Jangezoon, but he perceived listening to the community requirements almost as a burden and stated that it felt like a '*pinata politica*' (political trophy)[51] and benefit-sharing as a sort of 'blackmail' (*chantaje*). He felt had done his utmost to secure acceptance of the project.

The reasons for the lack of compliance with the essential requirements of engaging in proper consultation and carrying out prior social impact assessment was due to the companies' failure to acknowledge that the Ch'orti' possess a collective right as indigenous peoples to their territory, and natural resources.

We spoke to members of the civil society, who argued that the denial of indigenous peoples' rights stemmed from an endemic cultural, juridical, structural and economic racism deeply rooted within the country. At a state level this was reflected in the budget and public policies.

During our dialogue with the companies' CEO, he admitted that he was not aware of standards enshrined in the UN 'Guiding Principles on Business and Human Rights'.[52] He accepted that he was not familiar with principles of corporate responsibility to respect human rights and therefore had not taken them into account when developing his business model.

In view of these shortcomings, it was unsurprising that the project had met with opposition from indigenous communities. Contrary to good business practice, the companies had no risk management assessment concerning opposition to the projects, despite the well-known fact there had been a history of protest in Guatemala concerning hydroelectric projects. The companies' lack of awareness of such responsibility, blaming opposition on the 'nature' of the Ch'orti's, who were described pejoratively as 'tribes [who] like to fight', was evident.

This was a poor business model, and expensive for the company. The CEO of both companies indicated that, in ten years, the companies had made no profits, and had faced only costs. 'It has been an arduous process. The road has [been] time-consuming, exhausting,' we were told. Investments would have been more wisely made had the businesses been aware of indigenous peoples' rights, on the one hand, as well as current best practices of a corporation's respect for human rights, on the other. There is an economic cost to disregarding the law.

Two visions of development

The arrival of the companies saddens us.[53]
They say it is to 'develop us'.[54]

They call it development, they bulldoze you over
and call it progress.[55]

They tell us, 'you're worthless, you don't know
how to make the most of progress'.[56]
Ch'orti' communities

On closer inspection, at the heart of the community divisions and the clash with the company, there were two different understandings of 'development'. While the company stated that 'there is no elegance in poverty'[57] and there is a 'reluctance to improve, reluctance to work'[58] among the locals, the Ch'orti's questioned the concept of development advanced by the company. They did not believe that the dam would lead to the development of the community. A villager from Matasano said: 'They call it development; the community calls it disaster.'

The community and the companies were, literally, not speaking the same language. One indigenous leader explained: 'They are right, the word "development" doesn't exist in any Mayan language. Mineral extraction, selling, having paper money in the bank is not our way of understanding life.'[59]

True development for the Ch'orti' means ensuring conditions for a *Buen Vivir* (Good Way of Life), which entails protecting the water, the forests, 'everything that has life', 'for those to come' ('*para los que vienen*'); the guarantee for those yet to be born.

The communities we spoke to wanted to see a form of development that would increase their ability to securely meet their basic needs while maintaining their traditional way of life. Members of Nuevo Día argued that what development means for them is: strengthening communal authorities and leadership structures; establishing public policies, which could make agriculture more productive and resilient; promoting ecological agriculture, both as a means of sustenance and an economic activity; ensuring the right to life and subsistence of the community within the natural ecosystem of which they feel they are the guardians. They said: 'Rivers represent life and spirit rather than usable energy. If you cut the river into two with a dam, that energy is lost.'

The dark business of light

Upon reflecting and talking to members of the civil society, UN mission and the Ch'orti' themselves, I had begun to understand why, despite all the above, companies were adamant about the dam projects.

In a study called 'El oscuro negocio de la luz' ('The dark business of light') by FLACSO, it was pointed out, as regards the electricity business in Guatemala: 'Guatemala has the highest electricity rates in Central America, the demand for service almost matched the offer; power generation mainly depends on hydrocarbons causing high levels of pollution; the costs are socialised and profits are privatised; and the legal framework governing the process, from power generation to commercialisation, lacks transparency and allows a series of privileges for which consumers must pay.'[60]

This made me feel rather uncomfortable. The Ch'orti' knew that any electricity project was not to benefit them.

The Ch'orti's resistance to the dam projects was particularly acute because it was seen as the thin end of the wedge. Several members mentioned the 'technological corridor' (*corredor tecnológico*) connecting the Atlantic and Pacific Oceans. It would comprise a road, railway, and oil and gas pipelines. This corridor would run through their communities in Chiquimula. They were aware of certain additional mining plans for the region as well.

The creation of additional sources of electricity would therefore only be the first step in the 'development' of the region, preceding the construction of massive infrastructure and the growth of the extractive industries. The Ch'orti' could see little promise for their traditional way of life and their ancestral lands should these projects come to pass. They saw themselves as taking a stand now as defenders of the land.

The traditional way of life of the Ch'orti' was agricultural. Indeed, Guatemala is a predominantly agricultural country with half of the population living in rural areas, most of which are indigenous.[61] The predominantly rural character of Guatemala means that large numbers of people continue to be dependent on the land for subsistence and survival. In 2014, serious food shortages were experienced in the area of Chiquimula and Zacapa (the area we visited), the so-called 'dry corridor', following a severe drought that placed more than a quarter of a million families in a situation of food insecurity. Corn production fell by 80 per cent and bean production by over 60 per cent. Despite emergency food supplies being ordered by Congress, more than 100,000 families were left without food.

Parallel to all this has been, paradoxically, the expansion of mining, oil and gas exploitation, hydroelectricity and other mega-projects. Such activities have had disproportionate impacts on indigenous peoples, giving rise to social conflicts and objections about the long-term damage to the environment and the capacity of local communities to maintain their subsistence way of life and undermining their right to food.[62] Since the 1990s, the government has taken a number of measures to facilitate foreign investment, and mining and oil and gas exploration and exploitation, including by requiring low or even no royalties from foreign companies. A subsidiary of the Canadian mining company Goldcorp, for example, was exempted from paying taxes on its Marlin mine.[63] The decision by the Pérez Molina government to raise royalties from 1 per cent to 10 per cent in December 2014 set off a barrage of complaints from foreign companies threatening to legally challenge the new law.[64]

A major development was indeed looming at the time of our visit: the Mesoamerican Integration and Development Project (MIDB) launched in 2009, with the aim of integration of the Central American region through a series of infrastructural projects. The project replaced the proposed Plan Puebla Panama (PPP), established to bring investment, regional integration and development through road, electrification, port and other construction activities. The PPP was vehemently opposed by civil society organisations as well as by the Zapatista movement in Chiapas, Mexico, claiming that the project favoured multinationals over local communities and the environment. Its replacement, the MIDB, includes the Interoceanic Corridor, which consists of a 370-kilometre-long and 100-metre-wide highway to link the Atlantic and Pacific Oceans and will include a four-lane highway and parallel rail, oil and gas pipelines. This is what the Ch'orti' referred to as the 'technological corridor'. This was the monster they feared. The then proposed corridor was due to be completed in 2020, and was intended to run through the Guatemalan provinces of Chiquimula and Zacapa, where the Ch'orti' people live.[65] As of spring 2024, the time of writing, it was still a project that faced wide opposition.

Consent

International law requires governments to consult with indigenous peoples prior to major and potentially disruptive development projects being approved and implemented.[66] This consultation should be 'in good faith with the indigenous peoples concerned through their own representative institutions, in order to obtain their free and informed consent prior to the approval of any project affecting their lands or territories and other resources, particularly in connection with the development, utilisation or exploitation of mineral, water or other resources'.[67] No project can go ahead legally without indigenous peoples' consent.

However, despite ratifying ILO Convention 169 in 1996, Guatemala had taken no steps to ensure that its laws and administrative regulations were put into practice. We had found that the prevailing legislation was inadequate and in open violation of the constitution and international undertakings of Guatemala. None of the officials who were interviewed by the fact-finding team seemed to be aware that consultation with indigenous peoples was an obligation when intended projects were to be built in indigenous peoples' lands, and that there were specific requirements to ensure that the consultation would be fair and in good faith and the indigenous people needed to give their consent. Indeed, the officials interviewed considered they were faithfully complying with existing consultative processes.

In fact, when asked whether they regarded the right to consultation of indigenous peoples an obligation on the part of the State on such projects in their lands, the officers of the ministry we talked to stated that there was no such obligation, and that, in any event, a 'yes' or 'no' from such communities was not binding in regards to the project. They also pointed out that there was a 'vacuum in the law', as no regulations concerning the right to consultation in the case of indigenous communities existed.

While there has not been a regulation specifically concerned with the right to consultation of indigenous peoples, it was not the case, as raised by members of civil society in Guatemala, that 'you cannot carry out the consultation because there are no regulations'. The Constitutional Court of Guatemala recognised, in a case concerning a project in San Juan Sacatepéquez back in 2009, that fragmentary as the existing infra-constitutional laws were, there were a number of mechanisms that could and had to be used in conformity to the undertakings of Guatemala on the right to consultation. In that case, the court recognised that the right to consultation, as reflected in the international conventions ratified by Guatemala, 'forms part of the constitutional block or constitutional corpus' of Guatemala and, as such, Guatemala had to guarantee its efficacy in all cases in which it was relevant.[68]

The Ch'orti's resistance

'It's grotesque what they do here,'[69] we were told by a member of a civil organisation. 'Companies are untouchable'[70] and 'people are intimidated, or the penal system is used in a spurious manner with people being accused of certain types of criminal offences such as illicit association, terrorism, kidnapping and then being remanded in custody without considering alternatives to custodial measures'.[71]

Resistance and protest have been part of the Ch'orti' community's response to the perceived threats to its ancestral lands and breaches of their national and international rights. They had expressed their opposition to the form of development on offer via the hydroelectric projects. This has escalated into a conflict between the community, local authorities and company representatives.

Activities of resistance and protest have been criminalised in a manner that was not warranted: the response of the state to the opposition of indigenous communities to the hydroelectric projects has been simply to repress such opposition using the criminal justice system.

We received information that community members had been charged with inappropriate offences, which had the effect of stigmatising and persecuting the community, and quelling the free expression of their views and will. The criminal justice system was being used to repress opposition. 'They blame our communities for the conflict,' we were told. 'We suffer discrimination from local authorities', and 'they make us out to be a group of gangsters'. Leaders felt smeared, criminalised and they feared for their lives. They told us: 'We've been called a "red alert zone". It's not true. We are defending our territory and the livelihood and natural resources of this community.'

On 18 September 2014, 200 agents of a special anti-riot police unit (PNC) were about to repress a protest using firearms, tear gas and anti-riot batons. On the other side of the Jupilingo Bridge, there were the Ch'orti's, with their children, demonstrating. It was a protest of the Ch'orti's against the hydroelectric projects on the Jupilingo, which had been met with a strong repression.

Then, the police started shooting – with live ammunition.

Several members of the community were injured, including one person who was wounded by bullets fired by the police. Five members of the Ch'orti' community were detained for 'disturbing public order'. They were kept in detention for three days and eventually released without charge. Referring to that particular event, the UN High Commissioner for Human Rights noted in its report on Guatemala: 'The work of human rights defenders was also obstructed by *arrests* and criminal prosecution.'[72] For the criminal legal system in Guatemala, however, the protestors were in conflict with the law.

In August 2014, fourteen members of the Las Flores Indigenous Council were called to a mediation meeting (*junta conciliatoria*) in the Chiquimula District Attorney's office (Fiscalía Distrital). Having received no prior information about the accusation, the leaders decided not to attend. They managed to find out that they had been accused of 'threatening behaviour'. The communities believed that it was part of a strategy to criminalise the community of Las Flores, a Mayan Ch'orti' village in the Municipality of Jocotán affected by one of the dams, which had just received formal recognition of its indigenous status.

We were told: 'Our lawyers are treated like terrorists.'[73] The International Commission of Jurists had raised a number of concerns with regard to access to justice for indigenous peoples, which go beyond linguistic barriers, although language is a major concern. These included the distance of the courts from the villages, making attendance at court hearings difficult and expensive, and the lack of access to legal representation. Persecutions, threats, arrests and even murders of indigenous peoples defending their right to the land have been reported to inter-governmental organs, such the Inter-American Commission on Human Rights.[74]

The report: Machiwar kayopá temeyum Iranon kuxpon tará

Once back in London, drafting the report with our findings was a painstaking process. It was tiring and, at points, appeared impossible to achieve within the constraints of my ongoing practice. It was a task that offered no reward beyond a job well done: establishing a legal truth, navigating dry law, collating the evidence, immersing oneself into the national legal framework and examining it against the yardstick of international law. I worked like a Trojan, tackling issues one by one, persistently, making a sustained effort. I had to go not only through a wide range of facts but through intricate national legislation, case-law covering an entire legal system and complex international case-law spanning decades (the sort of study academics would dive into for years), all in a matter of weeks. Mine was an endeavour to identify the shortcomings, the principles of the constitution, the gaps in the legislation to be corrected, the twists of the jurisprudence. I wanted to resolve the puzzle of what trumped what, of what the law was. This was the work of a technician of the law. I had a sense of responsibility, an urgent need to distil facts and reach findings and legal conclusions, and make justice of the evidence we had heard. All the years of practice and knowledge in the area came together to assist me in understanding the applicable law. I left no stone unturned.

By the time I completed the report, I had worked around 400 hours pro bono. I had done my job well but had no expectations that the report would be anything other than a starting point for understanding the conflict surrounding the case of the Ch'orti' against the hydroelectric companies. Little did I imagine the impact it would have.

The report meant a lot to the Ch'orti'. When a German parliamentarian visited Guatemala in 2017 and met with them, a community member of Nuevo Día stood up to address the parliamentarian, holding the report, and proudly announced: 'This is our bible, this says everything, all our rights are written in here', and solemnly he handed down the report we had prepared.

Time passed rapidly. One afternoon, in July 2017, I received unanticipated news. An Appeal Court of Guatemala had handed down a judgment ordering the restitution of their ancestral land to the Ch'orti'.[75] In arguing the case, the Ch'orti' had used our report to support the legal entitlements. When the judgment came out, a leader of the Ch'orti' said: 'The Ch'orti' people is alive' ('*El Pueblo Ch'orti' está vivo*'). The companies opposed the judgment, and the case was then elevated to the Constitutional Court, which confirmed the decision in favour of the rights of the Ch'orti' over their ancestral land.

The Ch'orti' had won.

The many hours I had spent sitting at my desk, drafting, analysing this case for the report, flashed through my mind. I was left with the realisation that, when one does a job well, no matter what the odds, it can be truly transformative.

FOUR

All Mankind is One:
The Camisea Gas Project in Peru and Non-contacted Tribal Peoples

When I saw that the Shakespearean actor Mark Rylance was going to re-enact the 1550 'Valladolid Debate' at the Middle Temple – my Inn of Court – I could not resist it. I had seen him in a number of Shakespeare's plays, including *Richard II*. But I had not imagined him re-enacting this sixteenth-century 'disputation' from the heyday of the Spanish Empire, where the rights and treatment of indigenous people in the Americas were discussed by two opposing sides. This had been the first time in history that a colonising nation had organised 'a formal inquiry into the justice of the methods used to extend its empire'.[1] On the one hand, Bartolomé de las Casas, a Dominican friar, arguing for the humanity and moral equality of the indigenous people in the Americas to the Europeans. On the other hand, Juan Ginés de Sepúlveda, arguing the opposite. Sepúlveda's arguments intended 'to stigmatise an entire race as inferior'.[2] This is a debate that I studied at school. To hear it re-enacted in the historical hall of the Middle Temple, where barristers train in the art of advocacy, felt special.

I sat in the first row. In awe, I watched Mark Rylance at his best, impersonating de las Casas, arguing that the indigenous peoples in the Americas had souls. He contradicted Sepúlveda's assertions that the Indians were barbarous, that they committed crimes against natural law, that they oppressed and killed innocent people, and that wars should be waged against infidels. Citing the Bible and canon law, de las Casas told Juan Ginés de Sepúlveda: 'All Mankind is one!'

At the end of the play, I went to greet Mark Rylance and even managed to get a picture taken with him. I told him I was one of them: a descendant of those indigenous peoples in the Americas, and now a barrister. The play had profoundly touched me. I also feared it remained terribly current in the twenty-first century.

I learned then that Rylance was a patron of Survival International, an international non-governmental organisation (NGO) whose focus was to advocate for tribal peoples' survival. When he heard I was a barrister, he immediately encouraged me to support Survival International too, and it was in this way that I ended up doing some advisory work for them. That ticket to a performance in Middle Temple would take me all the way to the Nahua-Nanti Reserve in the Peruvian Amazon.

The Camisea Gas Project

The Nahua-Nanti Reserve, located 100 kilometres from Cuzco in Peru, is home to a number of indigenous Amazonian tribes living in voluntary isolation.[3] It is an Amazonian reserve comprising 450,000 hectares, in South Eastern Peru, established to protect uncontacted indigenous tribes living in voluntary isolation.[4] NGOs note that it is 'the buffer zone to the Manu National Park, one of the world's most important protected areas'[5] and a UNESCO World Heritage Site. The combined protected spaces represent 'one of world's most valuable biological reserves'.[6]

I'm not sure why I'm so fixated by peoples like these last tribes in the world, in their voluntary isolation. It is not just the wisdom that they own, their knowledge of the Earth, their minds untouched by the modern world, their capacity to see what we no longer can or have forgotten how to perceive that I am drawn to. It is the strong sense that, behind the increasing physical destruction of their ancestral land, the rainforest, there is fundamentally the same despicable rationale that made Sepúlveda call the indigenous peoples from the Americas '*homonculi*', i.e. no people.[7] It is a sense of justice.

Indigenous peoples in 'voluntary isolation' are indigenous peoples or segments of indigenous peoples who do not maintain sustained contacts with the non-indigenous population, and who generally reject any type of contact with persons not part of their own people.[8] They may also be 'peoples or segments of peoples previously contacted and who, after intermittent contact with the non-indigenous societies, have returned to a situation of isolation and break the relations of contact that they may have had with those societies'.[9] They are 'the last peoples who were not colonised and who do not have permanent relations with today's predominant national societies'.[10] Recognising that these peoples live in a unique situation of vulnerability,[11] the Inter-American Commission for Human Rights has acknowledged that 'these peoples and their ancestors have lived in the Americas since long before current States came into existence'[12] and that '[t]oday, very few of them survive, and many are at risk of disappearing entirely'.[13]

Essential to the survival of peoples in voluntary isolation are territorial reserves (or *Reservas Territoriales*) such as the Nahua-Nanti. But the Nahua-Nanti was under threat, from an oil company working inside the Reserve. The Camisea Gas Project is 'one of Peru's largest hydrocarbon discoveries'[14] and involves the exploration of gas in one of the largest undeveloped gas reserves in South America. This project (which covers concessions in segments of land known as 'block 88' and 'block 56') is being carried out in the southeast of the Peruvian Amazon[15] by a consortium of gas companies led by the Argentine firm Pluspetrol and including the US's Hunt Oil, Spain's Repsol and Sonatech Peru Corporation.[16]

In the Amazon

'One thing I'm particularly keen to get a better understanding of is the legal status of Reservas Territoriales in Peru,' said Rebecca, a caseworker at Survival International. 'My understanding is that they are intangible, but I would like to know whether it is illegal or not for oil and gas companies to work inside these reserves.' At that time, Pluspetrol was indeed working inside the Reserve.

Up to that point I had never worked on a case concerning indigenous peoples in voluntary isolation. I had just qualified as a barrister in England, and a colleague, a fellow barrister whose office was next to mine, joined me for lunch in my office. She must have been intrigued by my work. Littered about my desk there were a number of reports I was studying to better understand the situation. 'I am preparing a legal opinion on a case concerning the Camisea Project and indigenous peoples in voluntary isolation,' I explained.

It felt a little strange, I admit, to come into contact with the Peruvian Amazon again, in this remote manner, right from the heart of the legal quarter in London. The first time that I encountered Amazonian tribal people had been on a visit to Puerto Ocopa in 1983, just under 200 km from the Nahua-Nanti Reserve.

Verdant and orange – that is how I remember Puerto Ocopa. The colours of the palm trees and their orange fruits were so vivid, the sky so blue. My sister and I had arrived there on top of a load of pineapples, transported by a pick-up truck. We had left Lima via La Oroya, located at an altitude of 3,745 metres above sea level, and had by then been travelling possibly for about a month until reaching the rainforest via Junin. Young women did not travel alone in Peru, especially in those days. But we were curious about the world, and intrepid.

The villages we visited left an enduring memory. One morning, early at 8 a.m., on arrival in the town of Chanchamayo, I couldn't help noticing that the entire town smelled of coffee. It was the most wonderful smell, and, over a breakfast of fried egg and rice, I soon discovered that this coffee was the best I had ever tasted in my life.

Arriving in Puerto Ocopa felt like entering a completely different world. I did not expect to find palm trees with these vivid orange fruits right in the middle of Peru. At the time this area was remote. There were no hotels whatsoever. We had a guidebook that my sister had acquired somewhere and we were literally working it out as we went along.

A large, red-bricked Franciscan mission – an image right out of the Werner Herzog movie *Fitzcarraldo* – stood in the middle of all this, among the palm trees. We sought refuge, hoping to spend the night there and somehow, reluctantly, we were given lodging. We slept outside of the mission, in a tent provided by the Franciscans. It was from this tent that, for the first time, I heard the sounds of the rainforest by night. It was scary and intense to listen to that noisy wilderness beyond the canvas.

Next day, we saw about fifty Asháninka children, dressed in school uniforms, having breakfast in the mission. The chiefs of the Asháninka tribes had given their children to the mission so that they would gain an education, we were told. But I wondered whether these children were to lose their language and their culture.

When we made it to the Tambo River, we saw them. The first indigenous Amazonian people, with little to no contact with the Western world, that I had ever seen in my life. They were dressed in their traditional brown robes, *kushmas*. They did not speak any Spanish. They had come to exchange things by the river.

This was my first encounter with the last peoples who had never been colonised in Peru.

Colonising the rainforest

Bartolomé de las Casas reached Santo Domingo, in what is now Haiti, in 1502, a part of the largest fleet ever to leave Spain for the New World. He was then a member of the lay clergy, and only eighteen years old, but he eventually joined the Dominican order. Santo Domingo back then was 'like a continent squished into half an island'.[17] It contained 'lowland rainforests, cloud forests, pine forests, dry forests, mangroves, savannah, coastal lagoons, salt lakes, a rift valley, karst land formations and four mountain ranges'.[18]

In *A Short Account of the Destruction of the Indies*, Bartolomé de las Casas gave an account of what he witnessed. His was the first denunciation in history of the casual slaughter of thousands of people in the Americas, 'a remote, barely imaginable quarter of the globe'.[19] His writing was a petition for justice. He exposed the ways 'in which towns, provinces, and whole kingdoms have been entirely cleared of their native inhabitants'.[20] 'I saw with my own eyes how the Spaniards burned countless local inhabitants alive or hacked them to pieces or otherwise condemned [them] to a lifetime of captivity and slavery', wrote de las Casas in his *Short Account*.[21] The men, he added, died down the mines from overwork and starvation.[22] During the three or four months he was in Cuba, 'more than seven thousand children died of hunger, after their parents had been shipped off to the mines'.[23] The Spaniards had transformed 'a trading and evangelising mission [. . .] into a genocidal colonisation'.[24]

'Are these not men?' asked fellow Dominican Antonio Montesinos in a sermon in 1511. De las Casas, likewise, had fought against the suggestions that these indigenous people of the Americas were some species of sub-humans.

The reason, wrote de las Casas, why the Spaniards had 'murdered on such a vast scale and killed anyone and everyone in their way' was 'purely and simply greed'.[25] 'They have set out to line their pockets with gold and to amass private fortunes as quickly as possible' so that they could then assume a status 'quite at odds with that into which they were born'.[26]

The colonisers had arrived, in the words of Las Casas, 'to drench the Americas in human blood and to dispossess the people who are the natural masters and dwellers in those vast and marvellous kingdoms, killing a thousand million of them, and stealing treasures beyond compare'.[27] But the medieval Spaniards could not penetrate the Peruvian jungle, and so failed to colonise the tribal groups.

These groups' territory remained untouched back then. However, in my career I have learned of other forms of dispossession threatening the survival of the Amazonian people, like the Nahua-Nanti, now at risk due to the Camisea project.

Twenty-first century dispossession

The Camisea gas project overlapped with the reserve created for groups in isolation – including the Kugapakori, Nahua, Nandi and other indigenous peoples. In fact, more than 70 per cent of Block 88 (an operating oil and gas block in Peru) overlapped with the reserve created by the State in favour of groups in a situation of isolation or initial contact who belong to the Kugapakori (or Matsiguenka), Nahua, Nanti and other indigenous peoples.[28]

The parallels with the greed of those sixteenth-century colonists, who came to indigenous peoples' lands and dispossessed them to make a quick buck, could not go unnoticed. But the year was now 2003, and administrative legislation had been enacted in Peru granting legal status to the Territorial Reserve (*Reserva Territorial*) in favour of ethnic groups in a situation of voluntary isolation.[29] On 24 April 2006, this protection of the reserves of

Indigenous or Native Peoples in Isolation or Initial Contact, was elevated to the rank of law.[30] So, the detriment of the Nahua's territory was illegal.

Yet, in 2012, Peru's Energy and Mine Minister, under the administration of Ollanta Humala, announced the expansion of the Camisea project further into territory of the Nahua-Nanti Reserve. The outrage was instant. NGOs warned that this expansion could lead to the 'extermination' of isolated tribes.[31] In 2013 the United Nations' Committee on the Elimination of Racial Discrimination urged the Peruvian government to 'immediately suspend' the expansion of the Camisea project.[32] In a statement dated 7 February 2013, a grassroots organisation called AIDESEP warned that the 'expansion violates the isolated indigenous peoples' rights under Peruvian and international law and could lead to their extermination'.[33] On 31 January 2014, AIDESEP made the State, the companies and the Inter-American Development Bank (IDB) (which was financing the project) responsible for 'the death or illness that may affect the tribes who ancestrally inhabit the territorial reserve'.[34] A 2014 report by the Forest Peoples Programme warned about the vulnerability of these groups; their lack of immunity to viruses and other contagious diseases rendered them especially susceptible to epidemics, causing large numbers of them to die.[35] In May 1984 the Nahua population had been reduced by almost 50 per cent due to outbreaks of respiratory infections to which they had no immunity after contact with illegal loggers in their territory.[36]

'The Battle for the Nantis', a report published in 2014, could not have put it more clearly. It argued that what the Nahua population was facing were a number of 'strategies put into practice [. . .] in the attempt to extinguish [the Territorial Reserves]'. [37] This was because the Nahua 'occupy the segment closest to a deposit of great interest to the oil industry in connection with the expansion of the largest Peruvian Project of hydrocarbon extraction: the Camisea Project'.[38] The report denounced 'manoeuvres leading to genocide with all elements of this penal typification, including, in this case the cognitive element ("knowingly").'[39]

Drafting my legal advice

In March 2015, three years after the expansion of the Camisea project had been announced, I was asked for my legal opinion. I was asked to look at the legal protection provided for Indigenous Peoples in Voluntary Isolation and Initial Contact living in the Amazon in Peru. In particular, I was asked to address what the legal status of *Reservas Territoriales* (territorial reserves) were, with a focus on those living in Nahua-Nanti Reserve. In other words, I needed to determine whether it was legal or not for oil and gas companies to work inside the reserves. This included considerations not only of human rights, but also those relating to the fragile biodiversity in the relevant Amazon jungle basin and their protection under international law.

International law, my specialism, is particularly relevant here because international conventions are part of the Peruvian legal system.[40] In fact, international human rights treaties protecting the Nahua have the same rank as constitutional norms.[41] This means that, if Peruvian laws were a kind of Matryoshka doll, international law would be the last, all-encasing doll. According to a provision of the Peruvian constitution, individual rights and liberties enshrined in it ought to be interpreted in agreement with the Universal Declaration of Human Rights and international human rights treaties ratified by Peru.[42] The Nahua had individual human rights, and also collective rights. It was my task to tell *Survival* how all this applied to the Camisea project.

I looked through the Georgian bay window of my office in Essex Street. It was a sunny morning. The world of the Nahua seemed extremely remote. Yet their fight was my fight too. The Nahua are the traditional stewards of our Earth's respiratory system – and they've cared for this Earth's respiratory system successfully for millennia. Our fate, here in London, is intimately tied to theirs for climatic reasons.

I started to draft.

Law 28736

The *dramatis personae* in a case are individuals, people. But sometimes, an Act, a piece of legislation, may play a pivotal role: like an actor. In this case, it was Law 28736: the law that had been passed in 2006 for the Protection of Indigenous or Native Peoples in Isolation or Initial Contact in Peru. What was its significance? What were its effects?

I decided to concentrate on understanding what the legal situation was before and after this law. The Peruvian constitution recognised a collective right to property of indigenous peoples with regard to their ancestral lands and this was the legal entitlement of the Nahuas even prior to Law 28736. The constitution recognises 'legal existence' to such communities and acknowledges that ownership of their land is imprescriptible.[43] The constitution also provides for the respect of the 'cultural identity' of the indigenous communities.[44] In addition to such constitutional provision, indigenous rights to property are also protected under two key international conventions.

The Inter-American Court of Human Rights has addressed the rights of indigenous peoples in a number of cases. As seen before, the right to property under the American Convention has been construed to include a collective right to property; the right to property of indigenous peoples is protected even in cases where there has not been a legal delimitation, demarcation or titlement (*titulación*) of their property in the domestic system; and, in the case of indigenous peoples, the right of property has been held to cover both tangible (land, territory and natural resources found therein) as well as non-tangible elements, such as their customs and culture, acknowledged to be closely linked to the indigenous peoples' right to their traditional territory. This meant that the Nahuas were entitled to their ancestral land even if not demarcated, as their collective property.

The ILO Convention 169, the most important international instrument protecting indigenous peoples, similarly recognises rights to property of indigenous peoples over their lands and territories. The term 'lands' includes 'the concept of territories, which covers the total environment of the areas which the peoples concerned occupy or otherwise use'.[45] For example, it would protect the territory of indigenous nomadic people. The term 'lands' also recognises the indigenous peoples' rights 'of ownership and possession [. . .] over the lands which they traditionally occupy'.[46] The convention also provides for indigenous peoples' right to decide their own priorities for the process of development as it affects their lives, beliefs, institutions and spiritual well-being and the lands they occupy or otherwise use, and to exercise control, to the extent possible, over their own economic, social and cultural development.[47] In essence, it protects their own vision of life. There is also a duty of the State to 'take measures in co-operation with the peoples concerned, to protect and preserve the environment of the territories they inhabit'.[48] The State, in essence, can't impose a project on them.

In relation to natural resources, 'natural resources pertaining to the lands of indigenous peoples, "shall be specially safeguarded"'.[49]

These rights include the right of these peoples to participate in the use, management and conservations of these resources. The ILO Convention further provides that 'in cases in which the State retains the ownership of mineral or subsurface resources or rights to other resources pertaining to lands, governments shall establish or maintain procedures through which they shall consult these peoples, with a view to ascertain whether and to what degree their interests would be prejudiced, before undertaking or permitting any programmes for the exploration or exploitation of such resources pertaining to their lands.'[50]

ILO 169 does not address the specific situation of indigenous peoples in voluntary isolation, but bearing in mind the positive duties state parties to ILO 169 have undertaken in respect of indigenous peoples (including the duty to protect and preserve the environment of the territories they inhabit), the choice of no contact by indigenous peoples in voluntary isolation is to be regarded as their response to consultation.[51] In other words, by choosing to be in voluntary isolation, the Nahuas have said 'No' to any exploration in their territory.

An evolutive interpretation of the law (that is, an interpretation where a term is given a meaning that changes over time), required by international human rights law, means that these provisions should be construed taking into account more recent developments in the area of indigenous peoples' protection regime. Most crucially, the Inter-American Commission on Human Rights has pointed out:

> Respect for the human rights of the peoples in isolation and initial contact requires a framework fully respectful of their right to self-determination, the right to life and the right to physical, cultural, and mental integrity of the peoples and their members, the right to health, and their right to the lands,

> territories, and natural resources that they have occupied and used from ancestral times.[52]

Self-determination, in essence, entails the right of indigenous peoples to be able to make their own decisions about matters that affect their lives. The right of self-determination is contained in Article 1 of the International Covenant on Civil and Political Rights (ICCPR) and the International Covenant on Economic, Social and Cultural Rights (ICESCR), both ratified by Peru,[53] and is part of its domestic legal system.[54] In their exercise of their right to self-determination, the Nahua were entitled to make their own decisions about matters that affected their lives in their territories.

In the case of indigenous peoples, 'there is a direct relation between self-determination and land and resource rights' that takes on particular importance in the case of peoples in voluntary isolation or initial contact.[55] Thus, 'one of the fundamental premises of respect for the rights of indigenous peoples in voluntary isolation is the respect for their decision not to have contact and their choice to remain in isolation.'[56] The principle of no contact is, therefore, the expression of the right of indigenous peoples in voluntary isolation to self-determination.[57]

Guidelines for the protection of indigenous peoples in voluntary isolation and initial contact of the Amazon region issued by the UN Office of the High Commissioner for Human Rights (OHCHR) – the first to be issued by the UN addressing the rights of indigenous peoples living in voluntary isolation and initial contact – enshrine the principle that in the case of peoples in isolation, the right to prior consultation 'should be interpreted mindful of their decision to remain in isolation' and their decision 'not to use such mechanism of participation and consultation'.[58] This would impose a duty on the part of the State to respect no contact and therefore respect the territory, lands, natural resources and right to life of such communities, and as put in 2007 by the UN Permanent Forum on Indigenous Issues, to 'guarantee the inviolability of [indigenous peoples in voluntary isolation and initial contact] territories and natural resources.'[59]

If the law was so clear, even prior to Law 28736 arriving on the stage, how could so many injustices be committed even today, i.e. their dispossession?

Underlying all the above international law, there is ultimately a *pro homine* principle: the law ought to protect people. The disappearance of indigenous groups living in voluntary isolation in the Amazon region would signify 'a loss of the irreplaceable cultural heritage of the last indigenous groups that have maintained harmony with their surroundings, as well as their invaluable knowledge of biodiversity and forest management';[60] 'a loss to all humankind'.[61]

As admitted by the National Environment Commission of Peru, to give an example, from 1950 to 1957, a total of eleven indigenous peoples disappeared from the Amazon as a result of deprivation of their traditional lands and natural resources, as in the Huallaga basin with the construction of highways.[62] The peoples who went extinct were the Resígaro, Andoque, Panobo, Shetebo, Angotero, Omagua, Andoa, Aguano, Cholón, Munichi, and Taushiro.[63]

I saw a video in which Amadeo García, the last man to speak Taushiro in the Amazon, is interviewed by journalists from *The New York Times*.[64] 'I am Taushiro,' he said. He used to belong to a non-contacted tribe. Now, he was the only survivor. 'He is a lonely man,' said a friend, and added, 'one needs to have communication with others'. When he dies, the Taushiro language will die too. The commentator observed, however, '[f]or Amadeo the language died a long time ago, there is no need for language when you are alone'.

So, there was already a legal basis that protected these people. Law 28736 did not create new territorial rights; what Law 28736 did was simply to implement existing entitlements of indigenous peoples in voluntary isolation. Already, since 1978, five territorial areas protected to benefit indigenous peoples in isolation or initial contact had been established in Peru.[65] The *ratio* of Law 28736 was to convert those five territorial reserves into Indigenous Reserves to 'standardise restrictions on the entry to Indigenous Reserves, as well as the exceptions',[66] as the regimes existing before varied.

The UN has acknowledged that the creation of territorial reserves responds to the principle of respect and duty to guarantee the right of indigenous peoples in isolation to their lands, territories and natural resources. The UN Guidelines for the protection of indigenous peoples in voluntary isolation provide that states must delimit the areas occupied by the indigenous peoples in isolation and to which they have traditionally had access. These areas, according to the UN, must be declared of transitory intangibility (incapable of being touched) in favour of these peoples until they decide their situation voluntarily. If we were to apply this principle to the Nahua, it means that the territory is intangible. The UN also provides that, in the areas adjacent to these delimited areas, specific protection measures must be established in order to avoid accidental contact. It also prescribes that the delimitation, in accordance with international instruments, must be based on the concept of the use that they make of it, this concept being much broader than that of possession.[67]

Amazonian tribes don't stay in one single place. They move. This is their way to protect their territory and not deplete the land. I had first understood this concept of nomadism within a wide territory, by the Amazonian people, during my trip to Puerto Ocopa. Our trip on that occasion ended in the lowland port of Puerto

Bermudez, with the seemingly limitless Gran Pajonal forest extending from it. This was a hot, malarial area that smelled of oranges about to go bad. There were illegal loggers around and we spoke to some of them. They regarded the indigenous tribes as 'lazy', because they 'did not work' and were not interested in 'getting a job'. When they did, mainly the young males, they worked as labourers, but only for a short period, just enough to earn some money, and then they went back to the jungle. They simply did not conform to the occidental life. They did not need to accumulate possessions and did not need money to subsist. The illegal loggers could not understand why the indigenous people did not cultivate the land and stay somewhere fixed.

They did not do so because it would require them to cut down the trees. Besides, the soil is unsuitable for agriculture because the Amazon basin is hot and wet. Organic matter breaks down quickly at high temperatures, making the soil poor. So, the indigenous tribes only plant a bit of achiote and yucca, and then only what they need. They stay for a short time in a place, then they move on.

Is it possible to accept that there is more than one way to live?

I remember finding myself on top of fallen trees in an area of the jungle near Puerto Bermudez. You could not see the ground. The only way to move was walking along the narrow trunk of a fallen tree. An Asháninka-looking child was helping me. His ability to move through the trees was amazing. His big toes had separated from the rest, almost as I had seen in primates, allowing him to grip. That image of me, scared of slipping, while a child guided me by the hand through the jungle, has stayed with me always.

So, the creation of territorial reserves implemented the duty to guarantee the right of indigenous peoples to live in isolation in their lands, territories and natural resources, and it is reflected in international human rights treaties.

The object of Law 28736 was one of protection. It literally sets out to guarantee the right to life and health, safeguarding the existence and integrity of indigenous peoples in voluntary isolation or Initial Contact in the Peruvian Amazon.[68] A section of Law 28736 deals with 'intangibility of the Indigenous Reserves'.[69] It clearly prescribes that 'no rights shall be granted that imply the use of natural resources, except for those that are carried out by the peoples that inhabit them for subsistence purposes and those that allow their use through methods that do not affect the rights of indigenous peoples in a situation of isolation or in a situation of initial contact as long as the corresponding environmental study allows it'.[70] It also prescribes that, in the event of locating a natural resource that can be exploited, the exploitation of which is of public necessity for the State, it will proceed in accordance with the law.[71] And the law includes all those provisions that require the consent of the indigenous peoples, and the respect to their self-determination rights.

In addition, the framework established by a predecessor of Law 28736[72] simply prohibited the granting of new rights that imply extracting natural resources and did not include a public necessity exception.[73] While Law 28736 incorporated a public necessity exception, it had to be considered 'in accordance with the law'. Article 8 of Law 28736 itself stresses that indigenous peoples in Voluntary Isolation or Initial Contact in the Peruvian Amazon 'benefit from all the rights that the Constitution and the Law establish in favour of the Native Communities'.

What trumped what? 'In accordance with the law' should be given the meaning of 'lawfully', which encompasses not only the procedural rules but also the substantive ones to guarantee the rights of the peoples in isolation under the constitution and under international law. 'It will proceed in accordance with the law' necessarily implies, therefore, to respect the 'right to no contact' and as a consequence the acceptance of the voluntary decision of indigenous groups to remain in voluntary isolation in their lands.

The devil is in the detail

The devil, of course, is always in the detail. And so, Peru issued a 'regulation' of Law 28736 in 2007.[74] A bit like with George Orwell's *Animal Farm*, this regulation had the effect of 'rewriting' the law, impeding its proper effect.

Chapter II of said 'regulation' deals with intangibility of the reserves.[75] Article 35 refers to 'use of resources by public need' as an exception to intangibility. The wording gives the 'Vice Ministry of Interculturality' power to decide on the matter. The language of Article 35 is very permissive, as to giving the possibility of allowing the exploration of the natural resource if the Vice Ministry agrees with it.

I could not see how the regulation could at the same time be intended to 'protect indigenous peoples in voluntary isolation' while providing for 'public necessity' exploration of their natural resources in intangible land. These groups, after all, lack immunity to common diseases, and major projects of exploration such as gas exploration/exploitation would destroy natural habitats on which these groups depend for their subsistence. It is impossible to reconcile the permissive language of this regulation with higher norms. The 'regulation' goes further than Law 28736, which at least contained language that could be interpreted in accordance with constitutional protections, albeit with difficulty.

For these reasons I found that the regulation was unconstitutional. The regulation was a low-ranking norm: that is, in the hierarchy of laws, it was below legislation. Under Peruvian law, when there is a conflict between the constitution (or norms given constitutional rank) and a lower norm, the constitution prevails.[76] And the constitution affords a high level of protection to the uncontacted peoples and their territories. The regulation, the legal basis for any gas exploitation being carried out in the Reserve, was simply illegal.

I realised that the regulation had been issued during the administration of Alan García Pérez, reportedly one of the most corrupt presidents that Peru has ever had.[77] At the conclusion of his first presidency, García Pérez was accused and investigated for corruption and illicit enrichment. He managed to seek asylum and leave Peru. But time caught up with him. On 17 April 2019, he shot himself in the head while hiding in his bedroom as he was presented with an order of arrest relating to investigations for corruption and bribes received from Odebrecht, a construction company.[78] Alan García Pérez, who once confessed to a reporter that, in his life, 'money arrives alone',[79] could never explain the origin of his wealth, which had enabled him to buy properties in Paris, as well as several properties in luxury areas in Lima and in exclusive beach resorts elsewhere in Peru.[80]

As to whether it is legal or not for oil companies to work inside these reserves, it is irrelevant that the discovery of gas took place *before* the reserve was created, as the right to inviolability of territory of the indigenous peoples stemmed from customary law acknowledged by the constitution and international law (part of the domestic legal system in Peru), and precedes any creation of reserves. As a consequence, the exploration and exploitation in the reserve, in my view, had been illegal, because it infringes constitutional rights and constitutional rank-like entitlements under international law, of the Indigenous Peoples in Voluntary Isolation and Initial Contact.

A loss to all humankind

I sent my legal advice to *Survival International*, hoping that it would give them the clarity they needed in their advocacy on behalf of the protection of the Nahua reserves. *Survival*'s feedback on the usefulness of the advice was positive. They now had the clarity to elevate their campaign in favour of the Nahua.

However, a year later, shocking news of a mercury poisoning 'epidemic' among the Nahua tribe in the reserve came to my attention:

> Up to 80 per cent of a recently-contacted tribe in Peru have been poisoned with mercury, raising serious concerns for the future of the tribe. One child has already died displaying symptoms consistent with mercury poisoning. The source of the Nahua tribe's poisoning remains a mystery, but experts suspect Peru's massive Camisea gas project, which opened up the tribe's land in the 1980s, may be to blame. The project has recently been expanded further into the Nahua's territory, prompting fierce opposition by the tribe.[81]

Whereas the Camisea gas exploration of Block 88 began its operations in 2004 and they sought a dangerous expansion in 2012 (which led to my instruction), the gas had originally been discovered by Shell using seismic tests and exploratory drillings in the 1980s, which among other things indirectly led to the first sustained contact with the Nahua, causing the deaths of at least 42 per cent of them within a few months.[82] This was written on the wall. The expansion knew it.

The 1550 'disputation' of Bartolomé de las Casas and Juan Ginés de Sepúlveda on the 'subjugation of Indians', which I had watched on stage at Middle Temple, came back to my mind. Why, centuries later, are the indigenous and tribal peoples still treated as if they were less human or as if they had no rights? Who, I asked myself, is the 'barbarian'?

We have set up a world system that relies on fossil fuels. They are extracted to the detriment and even extinction of peoples like the Nahua. I was starting to see this clearly, and it made me feel uncomfortable, as I felt almost complicit. And then, sometime later, in 2017, I saw a piece of news that disturbed me even more. It was a report by the BBC. The piece pointed out that 'controversial gas from Peruvian Amazon [had] arriv[ed] in the UK'.[83] And added, 'it is thought to be the first shipment to the UK from the Camisea project in rainforest 60 miles from Machu Picchu'.[84] Was I relying on the Camisea gas to heat my flat that winter? If so, can I possibly call myself more developed than my ancestors?

Until I worked on this case, I had never asked myself where the gas I use in England came from. I must have been going through life like a zombie – without proper attention, without questioning. High in the Andes, people use dried cow dung to produce fire to cook. My ancestors certainly did. If moving away from gas implies exploring new systems and less consumption, I am prepared to do so. To go back to simplicity.

There is an assumption that realities like the Nahua in the Amazon are self-contained in some far-away lands. During my practice as a barrister, however, I have come to understand that this is not the case. When I learned that natural gas from the Nahua-Nanti Reserve was shipped to the UK, in a tanker owned by Shell, I realised that what was happening thousands of miles away from us could painfully implicate us.

Another piece of news, by Deutsche Welle, read: 'Amid a heated national debate over fracking, the UK has received its first natural gas cargo from the Peruvian Amazon. Critical voices, however, are getting louder. Is there no better alternative?'[85]

If Europeans were to be given an opportunity of rewriting history, back to the time of Columbus's voyage, with the knowledge of what the conquest of the New World did, would they do it again? This may be a second chance for humanity – to stop the destruction of entire peoples.

If there is something historically oppressed peoples have learned, it is resilience. Indigenous organisations in Peru have not stopped their opposition to Camisea. I know that they will persist, as they have since de las Casas first witnessed their plight. In 2021, Amazonian indigenous organisations appeared before the Constitutional Tribunal in Peru to ask the court to declare the intangibility of the area of Block 88 of Camisea to be free from any exploration or exploitation of hydrocarbons.[86] Their legal fight may have to go all the way to the Inter-American Court, but I am certain that they will prevail because the law is on their side.

FIVE

Sinking Islands: The Torres Strait Islanders

The auditorium, overlooking Hyde Park, was full. It was the morning of 6 September 2018 and I had taken the podium at WilmerHale, a leading solicitors' firm in London.

It was a fine morning and it should have been like many public speaking engagements I have attended, but this was different. I was going to go against the public view of one of the leading barristers in the country. Philippe Sands, KC, had posited three years before, in a public lecture entitled 'Climate Change and the Rule of Law' at the UK Supreme Court premises in London, that international courts of limited jurisdiction (which oversaw compliance with a specific international treaty such as international human rights courts), as opposed to courts of general jurisdiction (such as the International Court of Justice), were 'unlikely to contribute in a material way to a broader response to climate change challenges'.[1]

'I disagree,' I said, in clear terms. There was a silence in the room. 'It is my view that not only human rights courts and the International Tribunal for the Law of the Sea are equipped to adjudicate climate change claims under international human rights treaties and the Convention on the Law of the Sea respectively, but these courts will also play a key role in implementing the Paris Agreement, in doing so.'

I was at a conference on small states, and my presentation was entitled 'Melting Glaciers, Disappearing States, and Endangered Populations: International Dispute Resolution for Climate Change'.[2]

I was certain, fearless. The Convention on the Law of the Sea, adopted in 1982, is an international agreement by about 168 countries around the world that lays down a comprehensive regime of law for the world's oceans and seas, including establishing rules governing pollution and marine protection. In my view, greenhouse gas emissions, gases in the Earth's atmosphere that trap heat, amounted to 'pollution' within the meaning of the term in this convention.

There existed, up to that point, no climate change litigation in international courts. There was no thinking about how climate change might have engaged the Convention on the Law of the Sea and, indeed, international human rights conventions.[3] In essence, there was no 'blueprint' for successfully framing a climate change case before international courts. But I had been doing some serious thinking on this and my assertions were the product of two years of careful reflection and research. And I knew how.

The Paris Agreement[4] (another international agreement) adopted by 196 parties at the UN Climate Change Conference in Paris, on 12 December 2015, did not include an enforcement mechanism. But it had set out key obligations of states in respect of the climate.[5]

By the time the Paris Agreement entered into force, climate change was, however, having a swift and negative impact on the planet. In fact, sinking islands had become part of a dystopian reality faced by a world increasingly affected by climate breakdown, with 'many islands [. . .] slowly but surely being submerged'.[6] Yet in 2017, the idea that climate degradation could be subject to an international trial appeared distant or unlikely.[7]

In 2016, at a conference in Stockholm, I had been asked to consider whether climate change could be addressed using existing international legal mechanisms. I believed it could. Two years later, I was addressing the legal community in London with the reasons why. Norms that existed prior to the Paris Agreement could be used to address urgent questions about climate change.

In my presentation I explained how. I gave as an example the possible jurisdiction of the United Nations Human Rights Committee, a quasi-judicial mechanism that oversees compliance with one of the most universal human rights treaties, the International Covenant on Civil and Political Rights, which is legally binding, and explained why a climate case before this organ was feasible and how.

Nothing foretold the impact this speech was going to have on international law.

Present in the room were some lawyers from a climate non-governmental organisation (NGO), who approached me after my presentation, particularly interested in what I had just said. We had a meeting some days later and, soon after, they instructed me as counsel on what became the first international case on climate change, the Torres Strait Islanders case.

The Torres Strait region and its inhabitants

> 'We as a people are so connected to everything around us. The island is what makes us, it gives us our identity. We know everything about the environment on this island, the land, the sea, the plants, the winds, the stars, the seasons. The island makes us who we are. Our whole life comes from the island and the nature here, the environment. It is a spiritual connection. We know how to hunt and fish from this island – to survive here. We get that from generations of knowledge that been passed down to us. I know every species of plant, animal, wind on this island, the way the vegetation changes, what to harvest at different times of the year. That is the cultural inheritance we teach our children. It is so important to us, this strong spiritual connection to this island, our homeland.'

These are Keith Pabai's words. He's a Boigu Islander[8] and his words were my introduction to the Torres Strait culture and its inextricable link to land and sea.

I had locked myself up for ten days in my flat (and got an extension to work an additional seven days) to advise on the evidence and draft an application on behalf of inhabitants of the Torres Strait Islands, for human rights violations because of climate change, to submit before the United Nations Human Rights Committee. Based in Geneva, in the headquarters of the UN, this is a quasi-judicial body comprised of eighteen independent experts appointed by states that monitors implementation of the International Covenant on Civil and Political Rights, one of the main UN human rights treaties. I metaphorically immersed myself in the Torres Strait, through the statements of my clients, as well as through aerial maps and videos.

My first sight of the Torres Strait Islands was a bird's-eye view: green spots scattered through an expanse of deep emerald and lapis lazuli sea.

The Torres Strait is a belt of sea that lies between Papua New Guinea to the north and mainland Australia to the south. It joins the Arafura Sea to the west with the Coral Sea to the east. It is part of Australia. Names like 'Arafura Sea' and 'Coral Sea' spoke of remote, far-away places. A diversity of large ocean predators: sharks, marlin, tuna, swordfish and sailfish live in the Arafura Sea's waters – one of the world's biodiversity hot spots. Almost entirely open sea, the Coral Sea, for its part, contains unique and – to a great extent – undocumented biodiversity, such as precious corals and glass sponges.

The world's largest reef system is found in the Coral Sea. A series of plateaux and slopes etched by undersea canyons, separated by deep ocean trenches, characterises its geographical structure. A quarter of all marine species live in coral reefs. But with the warming of the water, primarily because of greenhouse gas emissions, the coral reefs are dying.

The significance of the Torres Strait for biodiversity and as a sensitive sea area was recognised in its designation as an extension of the Great Barrier Reef by the International Maritime Organisation.[9] 'The tidal influences of two ocean systems result in frequent anomalous tidal regimes and have a great effect on the area's biodiversity', the designation noted.[10] The massive freshwater and sediment input from nearby coastal rivers further influences this unique marine ecosystem. Benthic (of or occurring at the bottom of a body of water) communities, fish assemblages, seagrass coverage and coral communities in this region have all been well documented. The Torres Strait provides a critical habitat for many vulnerable or endangered species, including dugongs and green and flatback turtles, as well as supporting commercial fisheries for tiger and endeavour prawns, Spanish mackerel, tropical rock lobster, reef fish, pearl oysters, trochus and beche-de-mer. Coral reefs and clear waters support a rich fauna of reef fish, molluscs, echinoderms and crustaceans. Only eighteen islands are inhabited. The Torres Strait thus retains a high degree of natural and wilderness value.[11]

The islands where my clients live are Masig Island, Boigu, Warraber and Poruma. They are First Nation Australians. I found that the Torres Strait Islanders are a (small) numerical minority within Australia – about 0.14 per cent of the total population[12] – although they constitute almost the whole of the settled population of the Torres Strait region.

While there is an overarching commonality in the culture across the Torres Strait region, each island has a distinct culture. Some even have distinct languages. Both the Torres Strait Islanders as a group, and each island's community, constitute distinct and identifiable cultural minorities within Australian society.[13] There are two main indigenous languages spoken in the Torres Strait region, which are distinctive to the region: Kalaw Lagaw Ya (western and central region) and Meriam Mir (Mer and Erub Islands). In addition, many Torres Strait Islanders use a Creole language that is also unique to the region for daily communication. This Creole is often referred to as 'Broken'/'Brokan' (as in 'broken English'). The uniqueness of Torres Strait Islanders' culture is recognised both in Australia and internationally.[14]

My clients' accounts were very detailed. They explained that their culture was intimately tied to traditional means of subsistence, especially fishing and gathering living marine resources. They explained that living from the natural resources is an important part of the community's communal and spiritual life. In fact, as noted by anthropologists John Cordell and Judith Fitzpatrick, much like some of their Aborigine neighbours to the south on Cape York and the coastal Papuans to the north, the Torres Strait Islanders 'do not regard land and sea as separate spheres'.[15] Essential and distinctive features of their culture are bound to the land and sea territories of their particular islands and the seasonal rhythms and life cycles in the Torres Strait. To me, therefore, the application of the human rights treaty in this case could not stop with the land, it also had to cover the sea. Australian territory, constituted of maritime fauna and flora, on which the Torres Strait Islanders depended and was part of their ancestral territory, was being degraded rapidly because of climate change. The assessment of the violations – the exercise I was undertaking – was not restricted to a material world. The spiritual world formed part of their culture, and the sea was also a key part of that world. 'Belief in a marine afterworld is a Melanesian custom', I read.[16] There was also the deep attachment to the place where they were born and where their ancestors were buried. They tended the graves of their ancestors, who, for them, had a continuing presence as part of the community. In essence, the Torres Strait Islanders lived with their dead.

Of especial importance is the tombstone unveiling ceremony that completes a two-stage burial rite. It is a way of reaffirming the bond between the living and the departed. I was particularly struck by this. The intimate relationship of the living with their dead, present in the Torres Strait, was not unknown to me. I had experienced it in Latin America too. As Cançado Trindade, who sat as judge in the Americas for many years, hearing numerous cases that shed light on the intimate relationship between the living and their dead, put it: 'the oldest cultures [. . .] teach that our dead remain alive within us, and only die definitively the day we forget about them.'[17]

Impacts of climate change in the Torres Strait

I confess that, as someone born in a city, I hardly paid much attention to the seasons, or to the natural world around me, when growing up. I grew up in Lima, a grey city, where, as they say, it never rains, it only drizzles. There are only two main seasons: summer and a mild winter more akin to European autumn. It never snows. For that reason, I only properly discovered 'seasons' in Europe. I first experienced spring or winter, properly speaking, at twenty-six years of age, living in London.

To the Torres Strait Islanders, by contrast, seasons punctuate their lives in a significant way. Keith Pabai remarked, 'Seasons are really important to our culture because they tell us how we live and what we catch at different times. They link to the four winds that we have on the island, that come at different times of year. The winds bring different species for hunting, fishing, harvesting. They are like the patterns that shape life on Boigu.'[18] He then stated: 'Now . . . you cannot predict the seasons.'[19]

I think that it was this statement that encapsulated everything that was taking place in the Torres Strait and, all of a sudden, it made me realise the dimension of the crisis. I was breathless. It was, frankly, embarrassing that I was only learning the importance of seasons, of climate stability, now – with the Torres Strait Islanders as my teachers.

Ted Billy, of Warraber Island, for his part, noted: 'There [were] a lot of trees – coconut trees, fruit trees – that have all gone. Since I was young, a lot of garden land has been eroded away.' The erosion in Warraber Island is so bad that, according to Billy, 'The island has actually changed its position and shape because of the erosion. The island is spinning clockwise. The south part is moving toward the western side, and from the western side it moves to the northern side. It's like it actually turns. On the western side of the island we used to walk inside the scrub, but now there are no more bushes now where we used to walk. We used to go camping in some places on the western side of the island, *but the land is no longer there*.'[20]

I only made total sense of this description when I visited Sylt, an island in Germany. Sylt was also vulnerable to climate change.[21] Yet the sand in Sylt had been stabilised with marram grass and *rosa rugosa* to avoid erosion. Unlike Australia, Germany had taken action many decades ago, although it still faces the impact of climate change. In contrast, the situation was worsening in Warraber: 'I think the erosion is getting worse every year. When the wet season comes and we have cyclones, the soil is not only travelling around the island, it's getting washed off. This year is worse than last year, and I'm sure that next year will be worse still, especially in the southern and eastern sides of the island,' emphasised Billy.[22]

I did not want to start my draft with abstract notions of climate science. I needed to start with the facts at a palpable level. I needed to tell the story of the Torres Strait Islanders in the language of human rights. I watched hours of film recordings and examined photographic evidence on the effects of climate change on my clients' islands. I read their statements and subjected them to a forensic analysis. I knew that nobody was going to read hundreds of pages of evidence if I did not take the trouble of going through each statement and distil the impact, heading by heading, of the changing climate on the Torres Strait.

Yessie Mosby, a Masigalgal man (originally from Masig Island), described the island, his home, as 'a small low-lying coral cay'. It felt to my ear as a precious jewel in the middle of the sea. With a 'high' vulnerability rating to sea level rise and a 'low' response options rating (there are limited options to move to a safer site on the island), Masig Island is at high risk from sea level rise. As Mosby put it, 'a rise of 20 cm will start to cause significant tidal inundation of the community'.

While Mosby's grandfathers were able to cultivate sweet potatoes, cassava, sugar cane, corn and watermelon, the land where that was possible has now been eroded. Trees (coconut, almond and other giant trees) have disappeared. In the past people 'lived off coconut',[23] which now is scarce. The rising sea level has made the coconut trees (on which the community relied for food) sick. He explained: 'We cannot eat the fruits of these trees because they do not have water in them. We cannot make coconut oil from them or use them for cooking. Others do not bear any fruit anymore. Here and all along this coast, you see dead trees.'[24] Further, the community in Masig is also experiencing the disappearance of water resources: wells that provided drinkable water for the community have now become 'brackish, contaminated by saltwater'.[25] The island is not only sinking but the foodstuff as well as the water resources of the island are being affected by climate change; and there are also big changes to the reef.

There are no longer the same amount of crabs. The ecosystem is changing (seagrass is disappearing) as some species the indigenous people use for their sustenance are becoming scarcer: for example, fish, octopus, spider shell or clamshell.

Nazareth Warria, also a native of Masig, described the importance of fishing to be able to fend for themselves. 'Fishing is important to save money. We catch snapper, coral trout, trevally.'[26]

I read in sources from the International Maritime Organization (IMO) that 'indigenous people of the Torres Strait traditionally hunt dugong and turtle and fish for a variety of marine species for food'[27] and '[t]he consumption of seafood by Torres Strait Islanders is amongst the highest in the world on a per capita basis.'[28] Yet all the species they know are fast disappearing.

Keith Pabai identified species of fish, such as barramundi, one of Australia's most iconic tropical fish, as being affected by ocean acidification. So is the dugong. 'There are definitely [fewer] fish than there used to be,' he noted. There are saltwater species that shouldn't be in the swamp, and others in creeks that should not be there. In addition, he noted, it's just 'getting hotter' on Boigu Island.[29] Weather changes were deeply upsetting 'when to do the planting' and 'when to do the fishing' in the islands.[30] According to Warria, droughts were 'taking away their livelihoods':

> I have noticed that the weather has changed. Before, around April or May the Sagerr wind would be blowing, but now we still get the Kuki, the monsoon wind. The monsoon is coming much later now. The monsoon rain should come in November or December, but now we still have that weather in March, which is a big change. I've also noticed that we have more weeds growing on the island, and different species of bird coming from the north, from PNG [Papua New Guinea].

> We've also suffered from no rain, and it ruins our garden. It has also been getting hotter. I feel that we're going to lose our vegetable gardens, and that's taking away our livelihood.[31]

Kabay Tamu of Warraber noted that 'the coral keeps bleaching and our reefs are dying. The fish are dying.'[32]

For people that, as stated by Stanley Marama, 'live by the seasons', such changes of weather due to climate change deeply upset their lives.

The Torres Strait Islanders also described the increased vulnerability of their islands to 'high spring tides and cyclones further south'. For example, between 19–21 March 2019, Masig was affected by strong winds and waves because of Tropical Cyclone Trevor, which crossed hundreds of kilometres south over the Cape York Peninsula. The sea level rise and erosion was happening fast and the communities in Torres Strait were in fear of forced displacement. For example, Masig had been losing 'about a metre of land per year'.[33]

The link with the Arctic

All of the above sounded horribly familiar to me, but from a reverse situation. It reminded me of the Inuit case,[34] which had been brought by Sheila Watt-Cloutier, an Inuk activist and petitioner who brought a case denouncing the effects of climate change in the Arctic in 2005. The Inuit had charged, in 2005, that their world was literally melting away. Watt-Cloutier denounced that 'The Arctic ice and snow, the frozen terrain, which we as Inuit have depended on for millennia, was [. . .] diminishing before our very eyes'.[35] This had been the first time that the effects of climate change had been framed as a human rights issue. The claim was against the United States. It failed. Prior to my being instructed in the Torres Strait Islanders case and to the

September 2018 conference in London in which I presented my position on the viability of climate litigation in international courts, I had studied the pleadings in this case carefully. I wanted to understand the gaps in the argumentation of the case, and reflect on how the arguments could be strengthened. I had studied those pleadings carefully, to identify their weaknesses so as to 'crack the code' to make a climate human rights case under a human rights treaty successful. It was with that previous understanding that I was now articulating the Torres Strait case.

The Inuit claim argued that 'nowhere on Earth has global warming had a more severe impact than the Arctic.'[36] It was alleged that global warming had already visibly transformed the Arctic, 'altering land conditions',[37] making the weather of the Arctic 'increasingly unpredictable' (with Inuit elders, who have long experience in reading the weather, reporting various changes in weather patterns in different areas of the Arctic).[38] 'Like the Torres Strait Islanders, by 2005 the Inuit were unable to predict the weather. They denounced decreas[ing] water levels in lakes and rivers, producing "changes in the location, characteristics and health of plant and animal species".'[39] They explained that natural drinking water sources had become scarcer and polluted. This was the result of the combined effect of the decrease in snowfall, permafrost melt, the sudden early melt, erosion, rising temperatures and changing winds.[40] Their submissions were detailed. Painfully so. Yet the Inter-American Commission rejected the claim without even considering it. Now, over ten years later, in a different forum, the climate was in court again. The international community had failed to address the issue back in 2005. Had it done so, we may have not be where we are today. This had to be corrected and I was determined to do my utmost to make that to happen.

Rising seas

The low-lying island communities of Boigu, Masig, Warraber and Poruma are among the most vulnerable populations in the world to climate change. They are now in the early stages of a slow-onset catastrophe. Paradoxically, like the Inuit, they are communities that have contributed the least to climate change. Sea level rise, extreme weather events such as cyclones and floods, unseasonable weather patterns, increased heat and salinisation of land were, in essence, threatening life on the islands. Threatening life at all levels. The rapid degradation of the environment was threatening the sustenance of human life in the Torres Strait and threatening other forms of life, such as marine life and flora. They had become sinking islands. The Torres Strait Islanders could not just up and move to mainland Australia. They are inextricably linked to their ancestral home.

Yessie Mosby put it as follows:

> I will probably be alive to see my children not have anything. When they are adults they will not have anything for their children. We will be living on another man's land [. . .] That is when my identity, the Masigilgal identity, will die. I know a lot to teach my children, but I cannot teach my children about their inheritance on another man's land. It won't have the sacredness and the power of our culture [. . .] Our island is the string connecting us to our culture. It ties us to who we are. If we were to have to move we would be like helium balloons disconnected from our culture. Our culture would become extinct. We would be a dying race of people.[41]

What Yessie was referring to was a right to life 'with dignity'. The notion of 'life with dignity' is familiar to me. It evolved in human rights doctrine from the Inter-American system, a system in which I litigated for many years. The right to life is not understood as imposing merely negative obligations on states (refrain from killing), but also positive obligations (to foster life), which entailed ensuring a core level of economic, social and cultural rights were realised. When analysing the dispossession of ancestral land of indigenous groups in the Americas, where they had practised their culture for centuries, the Inter-American Court had recognised that their right to life with dignity had been breached, because they could not fulfil their dignity as persons outside their ancestral land.

This is precisely what was happening with the Torres Strait Islanders. The changes observed were already affecting *ailan kastom*, the body of traditional customs, knowledge and practices observed by my clients and their communities. The climate change impacts on the marine environment were harming traditional ecological knowledge and important cultural practices. They described the bleaching and death of the coral reefs that are the foundation of the marine ecosystems on which their way of life depends. They related the collapse of turtle and dugong populations, which are important ceremonial foods and equally important as a source of cultural and spiritual identity. And they told of the loss of fish and other marine species, such as crayfish (tropical rock lobster), which are forecast to be negatively impacted by warming waters.[42]

Forced displacement was a real threat. Stanley Marama explained:

> If I was asked to leave this island, I can't. I can't leave my community, because this is my home . . . I can't leave my grandfathers, grandmothers, and other relatives behind in the cemetery. It will affect my family, my kids, my grandchildren, because we spend our whole lives in the community of Boigu. I think if we had to move down south, this is when we would

question ourselves. We can't fit into life down there, because a lot of things are connected to white man. If we had to leave behind the culture and leave behind the community, we would leave behind our history – our grandmothers and grandfathers. That will affect us, make us question who we are.[43]

Their sacred sites were in their ancestral islands, he added.

'Many sacred sites are in the village . . . on Boigu, some are in the swamp. It's important to visit the sacred sites on the island because that's where we can get the blessing. That's our culture and we believe in that. It's a reality, that's our life, where we find the extension of our history. We have to visit those sites to get that blessing, to give us the understanding of our life, to refresh our thoughts. It's where we practice sacred cultural events like initiation'.[44]

Ted Billy expressed himself likewise: 'If we have to move from this island, then as long as we live until we die there will always be an empty space in us – that is how much we are connected to the island, to the place that we live.'[45] He further explained: 'There are also several sacred sites on the island that are really important to us, we still have a real connection to the history of the site. That would be lost and missed if we had to move away. We won't be able to pass down to the younger generation the information about the stories and history, their culture.'[46]

Take the example of the dugong. The Torres Strait dugong and turtle fisheries are 'traditional subsistence fisheries limited to traditional Inhabitants of the Torres Strait' and recognises that 'hunting for dugong and turtle is an important part of the traditional way of life and livelihood of Torres Strait Islanders and is also a major source of protein in their diet. Dugong and turtle may only be taken in the course of traditional fishing and used for traditional purposes'.[47] Under the Native Title Act 1993, Traditional Owners (like my clients) 'have the right to take marine resources, including hunting of marine turtles for personal, domestic or non-commercial communal needs and in exercise and enjoyment of their native title rights and interests'.[48] In the Torres Strait, the turtle mating season is known as *surlal* or *surwal*. I learned that the first turtle to be caught in the *surlal* is generally female. 'Larger and fatter than their mating partner, the meat and oil they supplied could sustain many people,' Leah Lui-Chivizhe observes.[49] In Poruma, an 'old woman' acknowledges the first turtle by sprinkling fresh coconut over the body and pouring the milk into the mouth of the turtle.[50] Scholars note that 'when the first turtle of the season [is] caught, it [is] brought onto the beach and turned onto its back and the people dance[d] around it and sang'.[51] It is observed: '[t]his, [. . .] was 'an expression of joy mingled with thanks and earnest entreaty that plenty more turtle might be sent'.[52]

If they were to move to mainland, Ted Billy explained,

> [t]he access to the traditional food from the sea would mean that the ceremony would be different. It would be much harder to catch turtle in the proper way in Cairns. The freedoms that we have to express our culture and to access our marine resources wouldn't be available. We would miss the access to our traditional food from the sea. I don't know how we would be able to keep passing down the knowledge, the traditional knowledge that's been handed down by our great-grandfathers, like how to hunt, what to eat in what season and all these things. I mean if we have to leave the island and move down to Cairns, what cultural history are we going to pass down to our

> future generations? We would lose our cultural history if we had to leave the island.[53]

Finally, Kabay Tamu's words lingered in my thinking: 'For us to be forced to move because of the impact of climate change will be colonisation all over again. That's how colonisation started. You disconnect the people from their land and they don't practise their traditions anymore. They don't speak their language. Once we are disconnected from the land, we are disconnected from our culture, language and traditions. We will be climate change refugees in our own country.'[54]

Australia knew

What really astonished me, and was very peculiar about this case, was the Australian government knew all of the above. The Torres Strait Regional Authority (TSRA), an agency of the Commonwealth government, had produced a Climate Change Strategy for the region, which acknowledged that:

> [t]he extent of the predicted effects of climate change in the Torres Strait region, along with the geographic, ecological, social and cultural characteristics make Torres Strait communities amongst the most vulnerable in Australia.
>
> The effects of climate change threaten the islands themselves as well as marine and coastal ecosystems and resources, and therefore the life, livelihoods and unique culture of Torres Strait Islanders [. . .] If strong action is not taken to address these threats, there is the potential for climate change impacts in Torres Strait to create a human rights crisis.[55]

From the outset of the case, I had been interested in finding the official position of Australia relating to the Torres Strait climate effects. Any admission, any acknowledgement of harm. And I was right to search: the evidence was in Australia's own records. The State knew that the effects on traditional ways of life and culturally important living resources from climate change were severe.

Although this had all been described as leading to a 'human rights crisis', Australia had done nothing to provide adaptation, or mitigation efforts to reduce greenhouse gas emissions. If anything, Prime Minister Scott Morrison's administration had been encouraging coal mining and the fossil fuel industry. I had this vivid image of Scott Morrison (when Treasurer) going with a lump of coal to question time in parliament. 'This is coal,' he said triumphantly, brandishing the trophy, as if he'd just stumbled across an exotic species previously thought to be extinct.[56]

As a journalist noted, 'the coal was produced as a totem of how the government in Canberra was going to keep the lights on, and keep power prices low'.[57]

We filed the claim on 13 May 2019. That same year, vast wildfires affected large areas of the Australian countryside due to abnormally high temperatures – yet another deleterious effect of climate change.

A foremost case on the right to life and systemic questions of law

'I don't see how the right to life is engaged in the light of the lack of loss of life,' said someone to me at the time. It was clear to me, however, that this was a central aspect of the claim: the right to existence. The obligations of the State to ensure this existence, to make sure that the environment was not degraded by man-making processes. The law did not require anybody to die to make a finding against a state if the State endangers life by its actions, or indeed inaction. And by Australia's own admission, the situation in the Torres Strait was life threatening.

I had the realisation that, at the time this climate change claim was filed, there was a general perception that the Torres Strait Islanders case would not be successful. There had been no precedent. But this did not concern me. In my analysis, we had a strong case.

The idea that a treaty that did not contain the expression 'climate change' could be interpreted to draw binding obligations for states in relation to climate degradation was novel back then. When, in his public lecture in 2015, Philippe Sands KC had been sceptical of human rights courts asserting that they 'might only ever have a limited role . . . unlikely to contribute in a material way to a broader response to climate change challenges',[58] he had said so because he considered that 'the treatment of the subject' before such courts, in his view, '[would] invariably be limited to the application of a particular international convention'.[59]

But, in my view, he had removed the possibilities given by rules of interpretation. As the UN International Law Commission (a UN organ devoted to the codification of international law and its progressive development) Study Group had pointed out, a limited jurisdiction does not imply a limitation of the scope of the law applicable in the interpretation and application of a treaty. To me, the standards laid down in the Paris Agreement were relevant.

Nowhere in the International Covenant on Civil and Political Rights is the term 'climate change' mentioned, yet by means of the application of the principle of 'systemic integration', obligations contained in the Paris Agreement could be made justiciable.

In her book *Problems and Process: International Law and How we Use it*, former ICJ judge Rosalyn Higgins draws attention to the role of 'the making of legal choices' by the judges, in the task of telling what the law is.[60] But it is not only the judges who make legal choices. In arguing a case and presenting arguments, counsel makes legal choices and advises a client accordingly.

A claim under the International Covenant on Civil and Political Rights follows some precise rules. At the time, the UN Petitions and Enquiries section limited claims to fifty pages. This required legal choices to be made as to what to include, given the complexity of the case, in the communication. It was hard to keep such a complex case within fifty pages. Among such a variety of issues that needed to be raised, as counsel, I made sure that submissions on the principle of systemic interpretation were made in the case, from the outset.[61]

As part of the steps taken to ensure the case of my clients were adequately considered by the UN Human Rights Committee, I travelled to Geneva, to meet up with possible intervening parties as *amicus*, an impartial adviser to a court of law in a particular case. It was summer of 2019 and it was hot. I held a number of meetings with several possible intervening parties. These efforts ensured that the former and current Special Rapporteurs on environment and human rights filed an *amicus curiae*, as did Martin Scheinin, a professor and former member of the Human Rights Committee himself, on specific points of law that benefited from additional legal treatment.

Then there came a bit of surprising news. On 22 December 2019, some months after the claim was filed, Australia announced the provision of 25 million Australian dollars to go towards building seawalls, to provide adaptation funding to address the erosion in the Torres Strait.[62] Despite the Torres Strait Islanders having tried for a decade to get these measures adopted, it was only after the filing of the claim that Australia took this step. Was this a strategic step by Australia to show that the international case was therefore unnecessary? The intention did not matter. The sooner they could take adaptative steps, in my view, the better.

January 2020 arrived and we had not yet had a response from Australia to the claim. It must have been hard for my instructing party, the NGO, to explain to our clients about the slow progress of the UN machinery.

At last, on 29 May 2020, nearly a year after we had submitted the claim, Australia filed its thirty-three-page response: Australia denied any wrongdoing. To Australia, climate change was a 'global phenomenon' for which you could not hold a particular state responsible. It also argued that it was incorrect to interpret a human rights treaty in the manner the committee had been invited to. The Paris Agreement was irrelevant to this interpretation, in Australia's submissions. It also considered that the Torres Strait Islanders had invoked a mere risk that 'had not yet materialised', and were therefore not victims under the convention. In other words, no harm had actually occurred. It invited the Human Rights Committee to dismiss the case.

Needless to say, I studied these submissions very carefully. This was a sophisticated piece of argumentation, which one had to address point by point. And I did. Yet time was short. We were given a very short period to present our observations.

As an advocate in this type of litigation, you may get the submissions from the other party suddenly. Dealing with other cases scheduled in my diary and managing my time added to the stress and intensity. I wanted to try to do my best for all my clients. I recall thinking about the best structure to address the submissions Australia had made and working hard, until my fingertips ached as a result of typing so long, so fast. In litigation we often have to work against the clock. Deadlines are fixed and merciless. I confess I used everything I knew (or had learnt over twenty years of litigation and scholarly practice) to draft my response. I focused on questions of general international law and persisted with a forensic approach to the facts. But when I finished drafting, I felt confident. It is this legal intuition, probably emanating from legal instinct accumulated in years of experience. I knew we were going to win the case.

We filed the observations on 29 September 2020. The forensic effort to detail the harms supported by specific references to statements and evidence. The need to pick up the legal points that really matter and argue them following an irresistible, logical, structure. The need to state the law clearly, convincingly, leaving no doubt. Those were my efforts. Then I pretty much collapsed out of exhaustion. I needed to sleep badly, and I did so, non-stop, for hours.

Australia was given another opportunity to make further submissions, which it did inexplicably slowly, on 5 August 2021, almost a year later. The leniency reflected in this prolonged period to provide further comments contrasted with the urgency of the matter (and with the tight deadlines the claimants had to meet). The Torres Strait Islanders must have felt upset about the waiting period for these proceedings. But to them the whole process was important. They had the same procedural standing as the State. This was historic for them.

We did not submit further observations. All that we needed to say had been said, and we wanted a decision.

This decision, putting an end to this contentious case on climate inaction, between ordinary people and a sovereign state, was being made in Geneva, by the Human Rights Committee. It was thousands of miles away from Australia, but would have a great impact on the future of my clients' islands. We were expectant and hopeful.

The views of the UN Human Rights Committee

This legal fight had lasted for nearly four years. By now, the uncertainty of the outcome had built up. At last, on 27 September 2022, the Human Rights Committee delivered its decision (or 'Views', as they are technically known) on the case. I was not nervous. I had confidence in the correctness of the arguments I had advanced on behalf of my clients. But the world had not yet seen any international decision that made climate change justiciable in a human rights court. Would this case make legal history?

I opened the email containing the decision as fast as I could. I held my breadth. Then I saw it. Against all odds, we had won. The Committee had found that Australia had failed to adequately protect the indigenous Torres Strait Islanders against adverse impacts of climate change, in breach of the International Covenant on Civil and Political Rights. It ordered 'full reparation' for my clients. This was the first international decision in history to have found a state responsible for climate inaction. It was the first case in which systemic interpretation of a treaty in the climate change context had taken place. This was the precedent that set out a new moment for international law in addressing climate change.

Despite Australia's opposition, the Paris Agreement was found to be relevant to interpret human rights obligations of a state in relation to climate action. Up to that point the framing of climate change before bodies with specialised *ratione materiae* jurisdiction or limitations on applicable law had not happened. The case became a landmark. The approach taken for treaty interpretation set several ground-breaking precedents for international law and climate justice.[63] It has now become a blueprint for treaty interpretation in international climate change cases. The full realisation of the impact of this decision, of having carved the right arguments, was still felt by me, years later. This was a systemic change.

We did not win on everything, however. While the Committee had found violations of Article 17 (protection of home) and Article 27 (minorities cultural rights), the majority had not accepted a breach of the right to life because, for this majority, a term of ten or fifteen years (for the islands to sink) was not 'imminent' enough to constitute a threat to the right to life. I had argued on the notion of 'imminence' in the context of the right to life, long and deep, and to me this majority was wrong. An unprecedented number of dissenting opinions of the expert members of the Human Rights Committee, which actually found a violation of the right to life, were issued in the case.[64]

The law is often clarified in an incremental way. So, to me, what the Committee got wrong will be clarified through the case law, including in other courts. My task as an advocate is to contribute to that process. I hope that the International Court of Justice (or 'World Court', as it is often referred to) takes the opportunity to address this issue, in the form of an Advisory Opinion, as the climate has now also reached this forum. I am acting in that case for the Kingdom of Tonga, a small island state also much affected by climate change. As I write, early in 2024, I am preparing submissions on its behalf.

After the Human Rights Committee had delivered its decision, and the news had made legal history, I met Yessie Mosby, one of my clients, and his son, Genia, during a visit he made to London. In my world, the everyday contact with lay clients is taken by the professional client, in this case the NGO. So, as a barrister, I did not have everyday contact with Yessie, but I had met him via Zoom meetings before. This was the first time I was meeting him, and his son, in person.

We deservedly toasted with a glass of Negroni while Genia fought for the head of the fish (which in Peru is believed to be nutritious and good for the brain) during lunch. To me it was the meeting of two cultures from the Pacific. Maybe, in the past, our ancestors had crossed the Pacific and met. There was something truly refreshing about them. Through the testimonies and words of the Torres Strait Islanders, they always referred to 'We' (referring to their communities), and rarely to 'I'. They perceived well-being and the future as something inherently linked to the collective. Have we lost that way of thinking of ourselves?

SIX

West Timor:
The Montara Oil Spill Case

> I repeat myself to say, it is simply not tenable in the twenty-first century for the Australian Government to ignore [the effect of the Montara Oil Spill] because those who are feeling the effect of Montara are first, non-Australian and second, too poor to enforce action.
>
> Letter from Ward Keller, an Australian lawyer, re the Montara Oil Spill, on behalf of the West Timor communities, to Malcolm Turnbull, Prime Minister of Australia, 7 January 2016

> Some of the world's poorest people have been rendered into even greater poverty and desperation as a consequence of the Montara oil spill
>
> 'After the Spill, Investigating Australia's Montara oil disaster in Indonesia', Australian Lawyers Alliance, July 2015

In 2019 a letter arrived out of the blue at my chambers. It was addressed to me personally and it had been sent from Kupang, the capital of the Indonesian province of East Nusa Tenggara. It was written by Mr Ferdi Tanoni, the chairman of West Timor Care Foundation, and it was an enquiry about instructing me on behalf of West Timor communities seeking reparation for harm suffered as a consequence of the Montara oil spill, one of the worst offshore oil rig accidents in history.

Unlike some of the other cases I've worked on, this was a story that had already made headline news, so I was instantly able to look up the public information on the spill. On 21 August 2009, the

Montara rig had exploded, as a result of a well blowout during drilling operations. In the midst of the sea, 220 km from the Western Australian coast, a tongue of fire from the oil rig platform was surrounded by fumes. The rig continuing leaking until 3 November 2009. By the time the blowout was controlled, seventy-five days later, at least 23 million litres of crude oil had spilled into the ocean between Indonesia and Australia. Satellite and aerial images of the cumulative oil slick showed the extent of the impact. The oil had spread over more than 90,000 sq km. It was unbelievable to me that damage of such magnitude had faced no legal consequences by the point that letter arrived on my desk a decade later.

Eighty-one villages on Rote Island and around the city of Kupang on Timor had been hit hard. A seaweed farmer had discovered a dark sheen across the water, waxy yellow-grey blocks floating in the sea. The seaweed he and his family depended on for their livelihood turned white and died. Around 100,000 people depended on seaweed farming, and they were now affected by the contamination. As Tanoni explained: 'the seaweed businesses of all these farmers were destroyed.' His English was limited, with a strong Kupang-Malay accent, but he expressed himself clearly and he was to the point.

Ferdi Tanoni was a small businessman, respected by the communities in Kupang. Located in the southwestern part of Timor Island, Kupang had, in the past, been an important trading post during the Portuguese and Dutch colonial era. When the fishermen and seaweed farmers started to realise the dimension of the spill, they went to his house, looking to him for leadership and help. The oil had not only decimated the seaweed farming, it had destroyed fishing grounds, affecting all the fishermen in the area. This was a humanitarian disaster. Tanoni set up an organisation called West Timor Care Foundation to pursue justice for his community. He thought that his quest would take some weeks. By the time I met him, it had already been a decade, and had completely changed his life.

The company operating the rig denied that the spill had reached Indonesia. So, too, did Australia. Ferdi Tanoni became a member of the task force at national level (in which governmental entities were involved) to get redress for the pollution. Shockingly, ten years on, they were still trying to get a remedy. I was instructed to advise on avenues for reparation. Tanoni forwarded documentation and I got to work beginning to understand the dimension of what had happened.

One of the pieces of documentation Tanoni sent was a link to a documentary film, *A Crude Injustice*, by filmmaker Jane Hammond, who had prepared the documentary about the aftermath of the oil disaster with contemporary footage. I watched the film in chambers, the offices from where I practice as a barrister. In a typically corporate room, located in the heart of the legal quarter in London, near the Royal Courts of Justice and miles away from the disaster, I watched disturbing images of the aftermath of the spill. Approximately 318,000 litres of oil a day had poured into the Timor Sea for seventy-five days, right where turtles, whales, dolphins and seabirds live in abundance.

The neutral tone of the beige walls in my office help me to remain sane when confronted with such realities. The quiet surroundings help me avoid becoming paralysed by the weight of it all, and I use this serene professional space as a retreat, a space where I can think quietly and make things happen. Prior to starting my practice in this set of commercial barristers' chambers, which I joined in 2014, I had worked for many years in international litigation, mostly in rather distressing cases. One day during my training, in chambers, I found myself working on a case where the crux of the matter revolved around some evidentiary procedural rules. The case had its own twists and turns and sent me back to some commercial law cases of the eighteenth century. I felt light. I felt strangely happy. Then it hit me. There was no human suffering in those files. No trace of blood. And this relieved something in my brain. I realised then that I had been traumatised by the suffering of the world in all my previous cases. So much so that a commercial case in which the issues were merely technical in nature felt like an airy feather, a much needed change.

The distance between the oil spill and where I found myself learning of it helped me to concentrate on the legal issues, which we needed to do in order to fight impunity in this case.

The leak from the Montara wellhead platform had not only damaged the environment but the health and livelihoods of fishers and seaweed farmers of thirteen regencies in West Timor, Indonesia. Hammond's was a remarkable documentary, which captured and preserved important footage showing the extent of harm done to the health and livelihoods of Indonesian fishermen and their families, and to the pristine marine environment. Certain images go straight to the centre of the nervous system and stay there. The film left me breathless.

A trip to Jakarta

On 30 June 2019, I travelled to Jakarta to meet Ferdi Tanoni, the chairman of West Timor Care Foundation. Ferdi was a short man who spoke with determination. He had already been seeking remedies for the spill for a decade by the time we met, and seeing him standing there, in the lobby of the hotel where I was staying, discussing a meeting with the task force next day, I could not but admire his continued energy in spite of the long journey he had already been on. There had been some proceedings in Australia, tort in nature (a lawsuit for harms caused), but they had been addressed against the company operating the oil rig only, and covered harms to seaweed farmers in just two of the thirteen communities affected. 'People are still suffering, from islands as far afield as Sabu, East Flores, Lembata and Sumba,' he told me.

This was my first time in Jakarta and we were speaking in a restaurant opposite a main governmental building, right in the centre. Jakarta reminded me of Manila. I had seen shanty towns not far from the city centre and was learning that less than 50 per cent of Jakarta's residents have access to piped water. A cockroach started to walk across my foot.

I looked up at Ferdi. I could see there was a certain fatigue, a certain discontent and a certain firmness in his face. But he was determined.

The meeting the following day was productive. Having examined the evidence, we discussed possible avenues for redress and met with other officers of the task force. I was asked to deliver a written advice, within twenty-four hours, on all those possible legal avenues, which included the possibility of an inter-state case. I got to work immediately on my return to the hotel. I did some thinking and delivered seven pages of legal advice. The routes that the case might follow were sketched out. There was hope.

After my meetings I decided to take a break in Indonesia. I wanted to establish a connection with the place and asked Ferdi whether it was a good idea for me to visit Kupang, the affected area, while I was there. He said that he needed to prepare for such a trip and suggested leaving it for another occasion. Instead, I spent a few days on the Andaman Sea, switching off from everything else and learning to love the place. Not all seas are the same. Each has its own colour and smell. One has to get to know a specific sea and learn to respect it. The Andaman Sea occupies a significant part of the Indian Ocean and it is turquoise. I started to love the remoteness of it all, the endless blue skies and the serenity of its sunsets, the green hills, the crystal clear waters of the sea. The pink sands, I learn, are caused by microscopic animals called Foraminifera that produce red pigment on the coral reefs. I had never seen manta rays in the sea. I spotted a sea turtle that was swimming towards the infinite, unbothered by the boat, and barracuda fish sparkled in the transparent water.

When I was on my way back to London, the West Timorese communities confirmed my instructions to file a communication before the UN Special Procedures. They had not yet gained the green light to convert this claim into an inter-state one, but the claim in their own right, against Australia, would finally put this dispute on the international map. Up to that point Australia had simply ignored the West Timorese communities.

Risks of oil drilling at sea

Shortly after my visit to Indonesia I spent a couple of months in Hamburg, the seat for the Tribunal for the Law of the Sea. It was a kind of immersion training. I learned that, historically, the response to oil spills, in particular the approach to risk, had been reactive rather than proactive. Big disasters in the shipping industry happen before proper regulation and compensation mechanisms were established. I also learned that, in contrast to the transporting of fossil fuels by ships, where there existed mandatory insurance that would allow prompt mechanisms to secure compensation in case of accidents and oil spills, there was simply no similar mandatory insurance for oil rig platforms for harm caused by catastrophic oil spills.[1] At every turn, I uncovered protocols limited only to shipping; there was clearly a legal gap when it came to oil rigs.

For example, in the case of the shipping industry, Protection and Indemnity Insurance (P&I) clubs, mutual insurance associations, provide cover for open-ended risks that traditional insurers are reluctant to insure. Typical P&I cover includes risks of environmental damage, such as oil spills and pollution. There is no equivalent to cover similar risks arising in relation to spills from oil rig platforms.

The BP oil spill of 2010 in the Gulf of Mexico (also known as the Deepwater Horizon oil spill) triggered some critical thinking on the nature and scope of insurance covering losses caused by catastrophic environmental disasters such as oil spills.[2] 'A mismatch between the losses resulting from oil spills, the insurance available for the victims of spills, the [civil] liability of the parties responsible for losses caused by spills, and the insurance available to the parties who face such liability' was noted.[3]

In Hamburg I asked a representative of a P&I why enterprises involved in drilling were not made to purchase *ex ante* (before the event) liability insurance that would cover the actual risk of oil spill catastrophes, for say US$10 or 20 billion. His response was that it was probably because the insurance industry would not be able to supply this much coverage. In other words, the risk was too costly. Not surprisingly, in 2012, Lloyd's of London, the world's biggest insurance market, spoke up about huge potential environmental damage from oil drilling in the Arctic. A Lloyd's report noted the lack of a legal framework of international liability and compensation regime for oil spills if drilling were to take place there.[4] It believed cleaning up oil spills in the Arctic would present 'multiple obstacles, which together constitute a [. . .] hard-to-manage risk'.[5]

So, who pays for these risks? The average person exposed to a disaster certainly 'pays', as in the example of the Indonesian fishermen and seaweed farmers. Legal consequences and seeking remedies from the polluter are an uphill battle, however. We ordinary people cannot drive a car without insurance. Yet an oil company can operate oil rigs without insurance, in case of spills. And the reason there's no insurance is not because insurance isn't necessary. Rather, the truth of the matter is that the potential damage is so great that nobody will insure you. Just as an insurance company will refuse to insure a demonstrably dangerous driver, they refuse to insure oil rigs. The difference is, though, that without insurance it's against the law to drive. Somehow, without insurance, it's not against the law to drill.

In the case of the Deepwater Horizon drilling rig disaster, it was only after much litigation, and amid public and political outrage in the US, that BP eventually accepted responsibility. They paid out $69 billion in the US, some of which went to US fishermen and businesses.[6] BP denied that the oil reached Mexico,[7] but a 'quiet' settlement of $25.5 million with Mexico was also reached in relation to the spill. And yet, reportedly, not a single peso went to an affected Mexican.[8]

Disasters like the Gulf War Oil Spill, in which Kuwait's oil wells were torched by retreating Iraqi forces, were approached via international mechanisms for compensation. The United Nations Compensation Commission (UNCC) provided a legal process to remediate and seek compensation for damage arising from the Gulf War, including the oil spill.[9] The example of the Gulf War Oil Spill is instructive because it shows that international law does not condone the pollution of pristine environment and environmental harm. Prohibitions have to have legal consequences if breached.

To go back to the West Timorese example, international law prohibits significant transboundary harm, and causing such harm gives rise to a legal consequence to provide reparations. To avoid justice, the perpetrators of such harm relied on the fact that the victims were too poor to enforce action against them.

Working on the case

Between the end of October and 28 November 2019, I prepared the submissions I was going to file on behalf of the peoples of West Timor and East Nusa Tenggara who were affected by the Montara Oil Spill. A central point raised was that states (in this case Australia) have the responsibility to ensure that activities within their jurisdiction or control do not cause damage to the environment of other states or of areas beyond the limits of national jurisdiction. The West Timor and East Nusa Tenggara peoples' right to a healthy environment had been violated by the spill, with all the substantive elements that this right encompassed – including the right to life in dignity and the right to a safe climate, clean water, healthy food, non-toxic environment in which to live, work, study and play, and healthy biodiversity and ecosystems.[10]

The Special Rapporteur on the implications for human rights of the environmentally sound management and disposal of hazardous substances and wastes had pointed out in the past that the primary duty to prevent human rights violations rests with states, irrespective of the increasing recognition of the responsibilities of business enterprises and other non-state actors.[11] Therefore states may violate their obligations under international human rights law when they fail to take appropriate steps to prevent, redress and remedy harm caused by private actors, as had happened in this case.[12]

The company that operated the oilfield (a Thai state-owned entity) and Australia, the country in whose jurisdiction the oil disaster took place, had failed to redress the harm even ten years after the spill. Contrary to the 'polluter pays' principle and to the right of access to justice and effective remedy, ten years after the spill the West Timor People had not been compensated for the devastating effects the spill had inflicted on their communities, lives and livelihoods. The harm to their environment had not even been assessed by the polluter, contrary to the West Timor and East Nusa Tenggara peoples' right to information. Thousands of West Timor and East Nusa Tenggara people have struggled to make a living and educate their children following the devastation of their livelihoods. Their plight has simply been ignored by Australia.

In my submissions I explained that numerous efforts to engage Australia to address the grievances of the West Timor and East Nusa Tenggara communities had been ignored and that it was for that reason that they had no other option but to elevate the matter to the international level, seeking the UN intervention.[13]

The Montara Oil Spill

A lawyer never stays within the first level of information on something – the kind of information you read in a newspaper. You think that you have grasped the key facts of something conveyed in a couple of sentences, but it may just be generalities. A lawyer peels away the facts in a process similar to peeling an onion, until they reach the core. Each level appears similar, but it is deeper, possibly revealing more facts. So, over three weeks in chambers, I reviewed all the available documentation on the event.

The Montara Development Project was owned and operated by PTTEPAA, a subsidiary of the Thai company PTT Exploration and Production Public Company Limited. The Development is located in a remote area of the Timor Sea, approximately 254 km north-west of the Western Australian coast, and almost 700 km from Darwin.[14] The fact that it is located in a remote area is relevant to the response. The distance of the Montara Wellhead from Rote Island, Indonesia, is 250 km – was the transboundary harm alleged by the fishermen plausible? I looked at a map, and could see that the Montara Wellhead is found right in the middle of the sea separating Australia from Indonesia.

I wanted to know what exactly had happened. I found that, on 21 August 2009, a small 'burp' of oil and gas was reported as having escaped from the H1 Well at the Montara Wellhead Platform. While the initial 'burp' subsided, approximately two hours later the H1 Well kicked with such force that a column of oil, fluid and gas was expelled from the top of the well, through the hatch on the top deck of the Wellhead Platform, cascading into the sea,[15] causing at least 23 million litres of crude oil to spill into the ocean between Indonesia and Australia.

For a period of just over ten weeks, oil and gas continued to flow unabated into the Timor Sea.[16] As stated by the Report of the Montara Commission of Inquiry[17] (which was set up by the Australian government itself on 5 November 2009 to examine the cause of the spill), the magnitude of the uncontrolled release of oil and gas from the Montara Wellhead Platform was 'the worst of its kind in Australia's offshore petroleum industry history'.[18]

While reading this I had the vivid image of the pink sand and pristine oceans that I had come to know and love during my trip to Indonesia. It appeared incomprehensible that harm of such magnitude could happen with total impunity in the twenty-first century.

On 5 November 2009, the Australian federal government announced the Montara Commission of Inquiry: an inquiry to be held with nearly all the powers of a royal commission to examine the cause of the spill.

Although the Montara Commission of Inquiry acknowledged that 'evidence before the Inquiry indicated that hydrocarbons entered Indonesian and Timor Leste waters to a significant degree',[19] the terms of reference for the Commission of Inquiry did not include transboundary damage. It had a narrow scope. Furthermore, mere weeks were provided for submissions, with a deadline of 22 December 2009.[20] This made it difficult for my clients, the peoples of West Timor, to present evidence. However, I made a thorough examination of the commission's report. I took a forensic approach, going through the report page by page without cutting any corners. I was after admissions and findings.

A key conclusion of the commission of inquiry was that PTTEPAA, the Thai state-owned entity operating the oilfield, had not observed sensible oilfield practices at the Montara Oilfield. Major shortcomings in the company's procedures had been '*widespread and systematic*, directly leading to the Blowout'.[21] It took me weeks to extract this from thousands of pages, but it was there, clear as daylight. I suddenly got angry. Despite this finding, how had they managed to get away with it?

The Commission also held that 'the evidence before the inquiry repeatedly showed that risks were not recognised when they should have been, and not assessed properly when recognised.'[22] In essence, the Commission found that the blowout was not a reflection of one unfortunate incident, or bad luck. What happened with the H1 Well was an accident waiting to happen. The Commission found that the company's systems and processes were so deficient, and its key personnel so lacking in basic competence, that the blowout can properly be said to have been an accident waiting to happen.[23]

The report further noted that 'ensuring the integrity of oil and/or gas wells (that is, preventing blowouts) is a fundamental responsibility of companies involved in offshore petroleum exploration and production'.[24] It stated that 'in the petroleum industry, well integrity is ensured by always having built in redundancies (secondary barriers) to safeguard against a blowout. Unfortunately, in the H1 Well there were no tested and verified barriers in place at the time of the Blowout.'[25]

So, the company had failed. But what about its regulator? Should they not have been ensuring compliance with rules to avoid fatal accidents?

To my surprise the Commission had also made findings on the regulator – an Australian state agency. In fact, the Commission had found that this regulator had lacked due diligence. It appeared that nobody had complied with the industry's own standards reflected in the Offshore Petroleum Safety Regulation of June 2009, cited by the 2010 Report of the Montara Commission of Inquiry, which states:

> In a complex, high hazard industry such as offshore oil and gas, society expects a robust regulatory regime in which operators maintain safety to minimise the risk of a major accident event and regulators provide assurance that this is being done.[26]

Of course society expects a robust regulatory regime because of the risks involved!

The Australian commission found that the Australian authority to whom regulation was delegated, the Northern Territory Department of Resources (the NT DoR), 'was not a sufficiently diligent regulator [. . .] it also adopted a minimalist approach to its regulatory responsibilities'.[27]

The Commission further held: 'The manifest failures within PTTEPAA extended to the interactions that the company had with the regulator, the NT DoR which in the Inquiry's view, had become far too comfortable. The Inquiry is of the view that PTTEPAA engaged with the regulator as if it were a "soft touch".'[28]

The Commission concluded further: 'The Inquiry is of the view that nothing should detract from the primary responsibility of PTTEPAA to ensure well integrity. However, the Inquiry finds that the NT DoR's regulatory regime was totally inadequate, being little more than a "tick and flick" exercise.'[29]

The result of this soft touch, 'tick and flick' exercise was immeasurable harm to the people of West Timor.

The response of the company and Australia to the Montara Oil Spill

Once it was clear to me who was responsible for the oil spill, I was interested in how the state and PTTEPAA had handled the event.

Instead of taking charge of the incident, Australia had left PTTEPAA 'to its own devices'. This was not a small finding. Under international law, a state has the obligation to prevent trans-boundary harm. Yet Australia had simply left PTTEPAA in charge.

But this was not, unfortunately, all. There was more.

Following the spill, on 23 August 2009, the Australian Maritime Safety Authority (AMSA) commenced spraying dispersants onto the oil's surface. Australia sprayed 184,135 litres of dispersants onto the sea, with seven types of dispersants, some of which are known to be toxic and carcinogenic. Two of those dispersants, Corexit 9500 and Corexit 9527A, have since been found by scientists to amplify the toxicity of oil fifty-two times. Of the total dispersants used, only one is approved for future use in Australia.[30]

An Independent Report commissioned by WWF-Australia (which was filed before the Montara Commission of Inquiry) noted that the national policy in Australia provided for 'mechanical recovery, as preferred method of clean-up'.[31] It also observed that 'because dispersants are a much less expensive and less labour-intensive response option than mechanical recovery, it is relevant to consider whether dispersants were emphasised to minimise clean-up costs', in the case of the Montara Oil Spill.[32] Now the picture of why harmful dispersants had been preferred was emerging. The clean-up method (mechanical recovery) prescribed by Australia's own manuals had not been followed for one single reason: cost. It was 'cheaper' to use highly toxic (possibly carcinogenic) dispersants. But, of course, not really cheaper. The costs were simply 'paid' by the poor West Timorese. The remedy was even worse than the original harm.

Indeed, the option of using dispersants was noted to pose additional environmental toxicity. Coral reefs, referred to by the scientists as 'rain forests of the sea' for their 'seemingly endless biodiversity',[33] are in serious decline, threatened by climate change and other forms of pollution. Reportedly, 'a quarter of the world's coral species will vanish by 2050.'[34] The chemicals of the dispersants, more harmful than the spill itself, affected a vast area of wilderness and biodiversity in the ocean, including vulnerable corals. The exact magnitude of the pollution would probably never be known.

The decision to favour dispersants over mechanical recovery resulted in a higher volume of the oil remaining in the marine environment. Because dispersants were used so persistently during the response, very little of the oil released during the blowout was actually recovered.[35]

The effects on West Timor and East Nusa Tenggara people

In September 2009, days after the spill, coastal communities in Indonesia witnessed oil close to or on the shore. Since then, they have experienced significant losses in their seaweed farming, fishing and pearling industries. Some sinister effects on the environment and on their own health were also starting to show.

West Timor communities saw with their own eyes oil washing into seaweed farms, onto beaches, onto the hulls of boats, and fouling fishing grounds and trawler nets. Witnesses described coral turning white; the precious farmed seaweed turning yellow, then white, and falling off its ropes, destroyed.

Communities described the white 'sickness' that later appeared on the seaweed and that worsened with specific currents. The death of mangroves removed a crucial bulwark to the ocean and there was subsequent flooding of villages. Catastrophic incidents of dead marine life that were reported around the time of the Montara Oil Spill included fish, birds and sea mammals.

The aftermath of the oil spill was followed by a reduction in the catch (fish) from Kupang Harbour of around 80 per cent. Thousands of tonnes per month of the city's critical food supply had been lost. Seventy 'long liner' fishermen operating from Kupang pre-spill were reduced to around ten long liners operating today. If pre-spill there were around 150 fishing platforms off Kupang Harbour, only five remained. The catch of resilient bait fish harvested by these platforms dropped by 80 per cent and remained below 50 per cent. Twenty-one to twenty-two tuna boats and around seventy lampara boats reported radically reduced catches even when venturing out of traditional Kupang fishing grounds to avoid the worst pollution. All this immediately affected thousands of people.

The small, previously profitable, scampi trawling industry of Kupang was forced out of business. In October 2009 nets were so fouled by Montara oil and dispersants that it was impossible to haul them on board.

Seaweed, both natural and that established in aquaculture programmes, has been decimated throughout the region. A baseline indicator to the ecosystem, this has an effect upon the full range of marine life. Numerous other fishing communities across the region have suffered similar reductions in catch and production.

Some communities also experienced adverse impacts on public health, including skin conditions, such as rashes, pus-filled cysts and inexplicable bruising following exposure to the ocean. Incidences of food poisoning were also reported in the spill's aftermath.

This was all consistent with what Ferdi Tanoni had told me. 'Fishermen and seaweed farmers who are in daily contact with the sea suddenly died, and others have very strange diseases. Itching, vomiting, bleeding to death,' he said. And it is still unknown as to whether people will suffer long-term health issues associated with the Montara spill.

In a report released by US non-governmental organisations EarthJustice and Toxipedia, 'The Chaos of Clean Up', chemicals identified as being present in Corexit 9500 and 9527 (which were used in the Montara spill as dispersants) were cited to have a number of adverse impacts on human health. One chemical, although not registered as a carcinogen, 'should be handled as a carcinogen with extreme caution', according to the New Jersey Department of Health.[36] It is possible, therefore, that further long-term health effects (including cancer) could follow other effects already experienced in communities, and have the potential to impact upon the kidneys, liver and other major organs.[37]

The island of Rote was particularly affected, as were the islands of Landy and Usu. But the impact of the oil spill was not restricted. It spread to a huge geographical area. People from the villages of Oesapa, Tablolong, Lifuleo; the West Timorese regencies of Belu, South Central Timor, North Central Timor and the sub-district of Kefamenanu, started to see adverse effects of the oil spill.[38] Damage to the seaweed was reported on Semau Island, around three kilometres from Kupang, with a population of approximately 13,000 people. Damage was further reported in Lembata Island, 190 km north of Kupang, and in the Sabu Raijua Islands, which lie west of Rote Island across the Sabu Sea. The islands of Alor, Sumba and Flores may also have been affected.[39] There were also reports of widespread hunger, with deaths attributed to starvation in subsistence communities.

I cannot imagine how Ferdi must have felt when he was first visited by the fishermen and seaweed farmers in Kupang, who showed him the effects on their skin, and told him about the seaweed going white. I have no doubt that, if Ferdi had not become involved, denouncing these effects in West Timor, this tragedy would never have been known by any of us. Ferdi was a terrific advocate; he organised meetings, travelled to Jakarta and demanded a task force be formed by Indonesia to respond to the plight of its people. He denounced things in the media. He travelled to Australia and met Australian lawyers and parliamentarians. His life was suddenly dominated by this case. He insisted on telling me all about the drastic losses of income to community and government and how children of all ages had been removed from schools because their parents were unable to pay the small but compulsory fees. This suggested the oil spill constituted a generational harm.

Ferdi met a delegation of Australian lawyers (Australian Lawyers Alliance), who visited the affected communities in

Indonesia in 2013, and prepared a report, 'After the Spill', which recorded the effects of the spill on West Timor and East Nusa Tenggara People.[40]

In the aftermath of the spill, there was also an increase in the number of unemployed West Timor and East Nusa Tenggara fishermen and seaweed farmers, and many have been forcibly displaced. The city of Kupang has already emptied out, because there is not enough fish stock for the estimated 7,000–10,000 fishermen who had been living there. Thousands of fishermen migrated to other provinces. Having lost their livelihood, some of the seaweed farmers and fishermen changed their occupation and became labourers in road or building construction. The economic impact on their families has been devastating.

Indonesia's Centre for Energy and Environmental Studies has estimated that the economic loss to the fishing and seaweed industries in Nussa Tenggara Timur amounts to approximately AU$1.5 billion per year since 2009.[41] By 2019, when I prepared their claim, that was at least AU$15 billion.

In short, some of the world's poorest people have been reduced to even greater poverty and desperation as a consequence of the Montara Oil Spill. The actions and omissions of the Australian government has caused these communities to lose their livelihoods and has forced them to live in abject poverty.

In an effort to access justice, Ferdi spearheaded a class action case based on torts against the company in Australia's courts, but such a domestic claim had a narrow reach. It only related to the impact concerning seaweed farmers in two regencies out of thirteen: that is, Kupang and Rote Ndao. It also assessed the responsibility of the company only and not that of Australia. Nevertheless, it was an important step to seek justice for what had happened, which eventually moved forward because of two factors:

the persistence of Ferdi, and the engagement of third-party funding, which enabled the claimants to have access to a court.

Australia's denial of responsibility

No attempt was made by the Australian government to assist Indonesian communities and their government towards investigation, let alone remediation, financial assistance or justice. The Australian government's official position on Montara has been recorded to be that 'it is beyond the Australian government's jurisdiction to compel a title holder to perform any investigative or monitoring activities in the waters of another country.'[42] This stands in contrast to international obligations on transboundary harm. Ferdi's instructing me had allowed me to bring an analysis of breaches under international law into focus in the Montara Oil Spill.

The Australian government had not, at any stage, required that PTTEP Australasia take any action to ensure that Indonesia was not adversely affected by the spill. Instead, the Australian government had continued to assert that any negotiations by West Timor communities must be with PTTEP Australasia directly.

On 31 August 2012, PTTEP Australasia was charged with four breaches of the Offshore Petroleum and Greenhouse Gas Storage Act 2006 (Cth), paying a $510,000 fine to the Australian authorities. None of that money went to the communities in West Timor.[43]

Vocal Australian parliamentarians, such as Rachel Siewert, a senator for Western Australia, however, have acknowledged 'that the Montara spill occurred in Australian waters but flowed into Indonesian waters'.[44] She visited the affected communities in Kupang, West Timor, in February 2014, where she met Ferdi.[45]

In witnessing injustices, some people do speak up. Writing to Malcolm Turnbull, then Prime Minister of Australia, on 7 January 2016, Ward Keller, an Australian lawyer acting on behalf of the West Timor communities, stated:

> The Australian Government's lack of action in the face of significant and mounting evidence of devastation of the Nusa Tenggara Timur fisheries and damage to the livelihood of thousands of Nusa Tenggara Timur citizens can be viewed as nothing less than a national shame.
>
> [. . .]
>
> In my letter to Mr Abbott of 24 November 2014, I made a comparison to Montara with the well-publicised Deepwater Horizon spill in the Gulf of Mexico in 2010, seven months after Montara. BP, the polluter there, has already paid billions of dollars of compensation for that event; the BP spill was immediately subject of decisiveness on the part of the US president. Full monitoring was instigated and continues. I repeat myself to say, it is simply not tenable in the twenty-first century for the Australian Government to ignore a remarkably similar event because those who are feeling the effect of Montara are first, non-Australian and second, too poor to enforce action.[46]

No action was followed by Australia.

The right to a safe, clean, healthy and sustainable environment

Had this oil spill affected Australian communities, there would have been an outcry in Australia and around the world. But because the people affected were indigenous communities from small islands, their rights were ignored.

If you are a white person from the developed world, it is possible that you have never experienced this sense of being trampled on. But the people of West Timor and East Nusa Tenggara have most definitely had that feeling of being ignored by those responsible for the pollution affecting them. It generates a deep sense of impotence.

Yet these communities had a sense of dignity, and a sense of right and wrong. So Ferdi was determined to open up a new front, with my help, to make the injustice they were experiencing visible.

The West Timor and East Nusa Tenggara people claimed in the letter I prepared on their behalf, addressed to the UN, that the actions and inactions of Australia, as described above, violated their fundamental right to a healthy environment. It was the first time that a diagonal claim for human rights violations (that is, a claim against other than one's country of territorial jurisdiction) had been made before these institutions.

The right to a healthy environment has been referred to in recent jurisprudence in regional fora as a right with individual and collective connotations.[47] The critical link between human beings' realisation of basic rights and a healthy environment has been recognised in international treaties and instruments including the International Covenant on Civil and Political Rights.[48] Australia is a party to this treaty, which it ratified on 13 August 1980. Environmental degradation affects the effective enjoyment of human rights,[49] as the Montara Oil Spill painfully showed.

The West Timor and East Nusa Tenggara people submitted that there is a connection between a number of substantive human rights (including the rights to life, integrity and dignity), and, indeed, the right to a healthy environment, and the correlative obligations on the part of the State. They also claimed that they held procedural rights vis-à-vis Australia, such as access to information, public participation and access to justice.

The majority of the environmental obligations rest on a duty of due diligence on the part of the State.[50] Therefore, in order to respect and guarantee the right to life and integrity, states are under a duty to: prevent significant environmental damage, both inside and outside their territory; regulate, oversee and control the activities under their jurisdiction which may give rise to significant damage to the environment; carry out studies on environmental impact when there exists the risk of significant damage to the environment, draw up a contingency plan so as to have in place safety measures and procedures for minimising the possibility of major environmental accidents, and mitigating any significant environmental damage that would have ensued, even when this may have occurred in spite of preventive actions on the part of the State.

A state is also obliged to act in accordance with the precautionary principle, when faced with possible severe or irreversible damage to the environment, even in the absence of scientific certainty, to co-operate, in good faith, for the protection against damage to the environment and pursuant to that duty of co-operation, notify other states that may be potentially affected when they become aware that a planned activity under their jurisdiction could give rise to a risk of significant cross-border damage in cases of environmental emergencies, as well as consulting and negotiating, in good faith, with the states potentially affected by significant cross-border damage. A state has the obligation to guarantee the right to access to information relating to possible negative impact upon the environment and guarantee access to justice, with regard to state obligations for the protection of the environment.

None of the above were observed by Australia in the manner in which the events unfolded and their aftermath.

I completed my submissions in early December 2019 and will never forget Ferdi's reaction when he read them. He was pleased, empowered, jubilant. He felt vindicated by this clear statement of the facts and the wrongs that had befallen the Indonesian communities. So far there had been thousands of pages produced in relation to the Montara Oil Spill, but from that maze of information, he now had ten pages of distilled facts and legal analysis, which became a powerful tool in his own advocacy through the media and the understanding of the case. It was the first time that a thorough analysis of the legal issues in the case and the responsibilities of Australia had been laid down in black and white.

After completing my task, I was exhausted and needed some days off. But whenever I closed my eyes, I saw the image of oil on the sea, and tongues of fire licking above the surface of the water.

The right to redress

The West Timor and East Nusa Tenggara People had argued before the UN that they were entitled to reparation, including measures of satisfaction, compensation and guarantees of non-repetition for damage suffered as a consequence of the Montara Oil Spill, under international law.

Upon submitting the denunciation on behalf of the communities, I made oral submissions before the Special Rapporteur on toxics and human rights.[51]

On 11 March 2021, this UN Special Rapporteur, jointly with other Special Rapporteurs,[52] forwarded a Note Verbale to the Australian Government. This was historic. They noted that they expressed 'serious concern at the alleged damage to the environment and human rights of the affected communities and indigenous peoples in East Nusa Tenggara [. . .] and the failure to provide a remedy for the alleged harm resulting from the oil spill'. They also expressed concern that this event 'disproportionally affected populations in vulnerable situations who rely heavily on the natural resources in and around East Nusa Tanggara'. They crucially noted:

> Serious concerns are expressed over reports that your Excellency's Government is failing to meet its international and extraterritorial human rights obligations, to protect the aforementioned human rights and provide effective remedy.

The letter also requested specific information about the event and follow-up situation from Australia. This marked the first time in which Australia was required to give accounts about its handling of the spill to an international organ. For the peoples of East Nusa Tenggara, this was historic. On 10 May 2021, the Australian Permanent Mission replied. Australia deflected any responsibility, as if the spill had nothing to do with Australia. When asked to indicate measures taken by Australia to ensure the victims had access to a remedy, the letter stated that: 'In August 2016, a class action was launched in [Australia] by Indonesian seaweed farmers seeking compensation from PTTEP Australasia', and then remarked: 'Neither the Commonwealth nor Northern Territory governments were named as co-respondents in the action.' In other words, it washed its hands – it replied as if Australia had no responsibility whatsoever in what had happened. The fact that Australia had not been sued in an Australian court did not mean it had not violated the rights of the Indonesian fishermen and seaweed farmers. This was a cynical response.

As an international lawyer, you expect states to comply with the law. So, seeing this response from Australia, a functioning democracy, was unexpected and disappointing. These were no ordinary harms and the failures of Australia had already been found by Australia's own Commission of Enquiry. This response was callous. As a litigator, however, nothing surprises me. I have over twenty years of litigation experience under my belt. A litigator ought to have sharpness but also stamina.

The East Nusa Tenggara peoples are resilient. They asked their government to take up the matter, elevating it to an inter-state dispute. Only an inter-state case brought by Indonesia against Australia could seek redress for what these populations had endured with the spill. But the decision to activate those procedures is not in the hands of the East Nusa Tenggara people. Ferdi has not given up and keeps requesting this in his interactions with government officials within the Montara Oil Spill Task Force in Indonesia.

In the meantime, however, the UN intervention had an important effect. The UN also wrote to the company and asked for a response regarding the allegations. This put some pressure on the polluting company. The ugliness of the event was now known to mainstream UN channels, and this changed reality must have been considered by the company. As in a game of chess, the denunciation I prepared opened up several fronts. Finally, on 22 November 2022, Thailand's national upstream company, PTTEP, stopped contesting the claim by the Indonesian seaweed farmers in Australian courts, and reached an out-of-court settlement with them, agreeing to pay US$129 million in compensation for the oil spill that followed the 2009 Montara blowout off the shore of Australia.[53] This was an important vindication that the spill had affected Indonesian communities and they were finally going to receive reparation.

Ferdi's persistence had obtained this important win. The strategic move to file a denunciation before the UN had contributed to this outcome. With this, the impunity in the case has started to crumble.

SEVEN

The Heart of the Earth: La Línea Negra

It was a Friday evening in November 2019, and I had gone across the river to a bar overlooking the Houses of Parliament to meet Louise Winstanley from advocacy project ABColombia. While other people use Friday evenings to switch off from work, we were meeting to discuss an environmental case with which ABColombia needed assistance. There was a sense of urgency.

I had worked with Louise before. Through her role as Programme and Advocacy Manager of ABColombia, she represented a group of five leading UK and Irish organisations with programmes in Colombia. I had been instructed by her in another matter previously, but this was the first time that we had met in person. And it was the first time that I heard about La Línea Negra (Black Line).

'It is a sacred place for the Arhuaco, Wiwa, Kogi and Kakuamo peoples,' Louise told me. The names of these indigenous peoples and the name of their sacred place sounded out of this world to me, and yet so familiar. This was no ordinary case.

The four peoples from the Sierra Nevada de Santa Marta had survived the Spanish conquest up in the mountains, in isolation. They called themselves The Elder Brothers and their spiritual guides were called the Mamos. They possess an ancient system of knowledge that they call the Law of Origin, which, in essence, are 'the original instructions given on how to cohabit in a harmonious way with life on the planet'.[1] Trained since childhood in the ways of nature for guidance, for years these spiritual guides had been talking about 'unknown illnesses', climate change and shortage of water.[2] They believe that 'the destruction of Nature produces illness from the diseases that are seen today'.[3] They saw destruction and thought that humanity had to return to the Law of Origin.

That Friday marked the beginning of my involvement with their case. Louise wanted to instruct me to act in a foreign court, the Supreme Administrative Court of Colombia (Consejo de Estado), to prepare a legal intervention to protect the Sierra Nevada de Santa Marta, an area of global importance, also known by the Arhuaco, Wiwa, Kogi and Kankuamo peoples as 'The Heart of the Earth'.

Sierra Nevada de Santa Marta: 'The Heart of the Earth'

> We are one with the water, with the earth, with the air, with the sun, with the thoughts, with the heart, with the spirit, with the body. We are one with the plants, the animals, the minerals and the diversity of humanity . . . For eternity . . . the vision of over 100,000 people living in the Sierra Nevada de Santa Marta, the Heart of the World.
>
> Ati Quigua, Defend the Sacred, Colombia

The Sierra Nevada de Santa Marta is a unique pyramid-shaped mountain located in the extreme north of the Andes, in northern Colombia. The geology of this isolated range of mountains is fascinating. Recent research by Smithsonian scientists found that the diverse rock record exposed in Sierra Nevada 'rests on an ancient foundation that is more than 1 billion years old' and that, over the last 170 million years, the mountain had travelled from Peru to northern Colombia, to finally rotate in a clockwise direction so as to open up an entirely new geological basin.[4] Today, it is the world's highest coastal mountain range.[5]

There are four indigenous peoples (the Arhuacos, the Wiwas, the Kogis and the Kankuamos) living on its slopes. They number more than 90,000, and they are survivors. Having been brutally subjugated by the Spanish conquistadores during the sixteenth century, they endured forced displacement during the armed conflict in Colombia. In 2009, the Constitutional Court of Colombia found that they were in danger of cultural and physical extermination. Their survival is tied to the survival of their ancestral land.

The peak of the Sierra Nevada de Santa Marta is located at an altitude of about 5,000 metres. At its base, on the shores of the Caribbean, a dense rainforest covers the low plains. As the mountain gets higher, the jungle is transformed into an open savanna and cloud forests. All this is what the Arhuacos, Wiwas, Kogis and Kankuamos, the indigenous peoples of Sierra Nevada, call the heart of the world.

The Sacred Space encircled by The Black Line

In the belief system of the Arhuacos, Wiwas, Kogis and Kankuamos, the Sierra Nevada de Santa Marta (*U'munukunu*, for them), their ancestral land, is surrounded by an invisible 'Black Line' (La Línea Negra or *Séshizha*). This consists of invisible lines 'of origin' that radially connect sacred sites in this territory with the peak of Cerro Bolivar, which stands at an estimated height of 5,730 metres.

This is not a merely material realm or a geographical space. The Black Line or *Séshizha* represents the spiritual world and the black matter of the beginnings of the Earth. '*Shi*', which is the infinite thread that connects the spiritual and energetic force of the sacred sites, in Kogi means thread or connection, while '*zhiwa*' in Wiwa means water, interconnecting the different dimensions of the ancestral territory, as well as the veins in the body. The Black Line, therefore, is the connection of the material world with the spiritual principles of the origin of life.[6]

For the indigenous peoples of the Sierra Nevada, the Black Line is a spiritual fabric, which, like a spider's web, feeds, supports, connects and orders the physical space of the Sierra Nevada de Santa Marta, and from there it connects itself to the world.[7] The Kogis, Wiwas, Arhuacos and Kankuamos think that both the black space that we do not see (*Séshizha*) and the material world (*mama sushi*), which is perceptible by us, are part of a unity that contain and move the energetic interactions that give rise to life. The visible and the invisible are present in the sacred spaces.

In their cosmogony, in the beginning there was nothing. All was darkness, which they call *Se*. Depth. There was nothing at all.[8] It was from this darkness that life emerged. In an illuminating documentary on the Kogis, by Alan Ereira, this concept of the beginning of the universe is taken to a renowned astronomer, Richard Ellis, to see whether it makes sense as science. The Kogi Mamos are right there, in Mill Hill Observatory, where they have come on a once-in-a-lifetime visit to the UK to collect a gold thread. To Richard Ellis, this notion of darkness makes sense (to my surprise, and to the surprise of the journalist interviewing him). To the West, 'dark energy', as put by Richard Ellis, 'is a recent discovery.'[9] He explains: 'It is a property, probably of space, that we haven't yet understood, that makes the universe accelerate and it's a mystery at the moment.' 'Does it fill all space?' asks Ereira. 'It fills all space,' responds Ellis, and then he remarks: 'It could take another hundred years before we really finally understand what's going on.'[10]

So the Kogis' cosmogony grasps issues we are still to understand in the West. The Kogi Mamo points out: 'We were left objects, like maps, that show us where to connect with the darkness. These places, or nodes, are called *esuamas*.'[11] The astronomer in the documentary then remarks: 'And these nodes are intersections of the material world and the darkness, is that right?'[12] *Esuamas* (or 'place of authority' to the Kogi) are found in the sacred sites along the Black Line. In Ereira's 2011 documentary, together with the Kogis, he walked to each of these places in the Black Line.

The Black Line encircles a ring of 348 sacred sites in the ancestral territory of the Arhuacos, Wiwas, Kogis and Kankuamos, and contains wetlands, mangroves, tropical rainforest, dry forests, high mountain forests, moors and glaciers, over 700 bird species, forty-four critically endangered animal species (including the jaguar, tapir, harlequin toad, the blue-bearded helmetcrest, among others), and 3,000 plant species.[13] In 2013, a scientific study that identified the areas most critical to preventing extinctions of the world's mammals, birds and amphibians named the Sierra Nevada de Santa Marta as 'the most irreplaceable site in the world for threatened species'.[14] The Black Line also includes the marine space.

Since 1979, UNESCO has recognised key ecosystems as Biosphere Reserves, essential for preserving the planet's biodiversity, although this 1979 recognition only covered a fragment of the indigenous peoples' ancestral land. As accepted by scientists today, the Santa Marta mountains are considered 'the most important continental centre of endemism in the world because of the remarkable richness of unique flora and fauna found there.'[15] The complex topography, geological history, variation in climate and in ecosystems in Sierra Nevada de Santa Marta create the conditions that allow for so many endemic species to occur in a relatively small area.

For this indigenous groups of the Sierra Nevada, all knowledge and history of the Earth and life are encoded in rocks and stones. They are the ancestral maps of Nature. Rocks and stones function as the bones of the Earth, the structure that allows it to keep the natural elements in their order and stability.[16]

I also learned that, for the indigenous peoples of the Sierra Nevada de Santa Marta, the Earth is a single interconnected living organism.[17] Therefore, the sacred sites of the Sierra Nevada de Santa Marta are connected with sacred sites in other parts of the world, which jointly maintain the natural harmony and equilibrium, and the sustainability of life on this planet. The destruction of these sacred sites implies the weakening of the connected energy. The destruction of mountains because of the mining of gold, coal, or extraction of oil and gas; the damming of rivers – all are described by the people of the Sierra Nevada de Santa Marta as depletions that impact other spaces, altering the functionality of the cycle of water, air and natural flows.

Alan Ereira's documentary features a meeting between Professor Baillie from the Zoological Society of London and the Kogis to discuss connectivity, a fundamental principle of the cosmovision of the Sierra Nevada people. Professor Baillie admits that the West need to understand connectivity far better. 'Right now we have a very basic understanding of how things interact and affect each other and I believe that this is essential for our future security,' he remarks.

Take the example of water. The Sierra Nevada is key for water supply, and the interconnections with water cycles beyond the Sierra Nevada appear quite obvious to the indigenous peoples. It supplies water to the extensive region of northern Colombia and the Caribbean seas. It is made up of glaciers (once perpetual snow), more than 300 lagoons, swamps and marshes, and thirty main rivers distributed in three hydrographic basins that drain towards the Caribbean, towards the Cienaga Grande and towards the Cesar River. It supplies water to a population of approximately 1.5 million inhabitants.[18] It also directly supplies water to two Ramsar[19] sites: Cienaga Grande de Santa Marta and the Cenagoso de Zapatosa Complex – enormous water and biological complexes where both endemic and migratory species from many places on the planet converge and which are also essential for mitigating climate change, because they absorb large amounts of carbon, improve water quality, provide livelihoods and protect coastal communities from natural phenomena and rising sea levels.[20]

But this sacred equilibrium is being messed up.

The Arhuaco, Kogi, Wiwa and Kankuamo also believe that Nature feeds on the actions and thoughts of people and that, if the territory is disordered and sick, it is because people have not known how to maintain a balanced relationship with it and that we humans damage its order with our own disorder. For example, when the rivers are diverted or the mountains are destroyed, to remove stones or lime.[21]

In 2018, La Linea Negra was finally recognised as subject of protection by Colombian president Juan Manuel Santos in a key piece of legislation called Decree 1500. But in 2019 a challenge to that decree in the highest administrative court, Consejo de Estado, aimed to quash it. The challenge, brought by the Colombian lawyer Yefferson Mauricio Dueñas Gómez, alleged, among other things, that protection of the area by Decree 1500 is against 'development'. The Supreme Administrative Court of Colombia is to determine whether Decree 1500, which protects this ancestral, sacred land, is to be annulled or whether it should be maintained.

A Kafkaesque scenario

A few weeks after my meeting with Louise, she forwarded the background information and organised a conference with the non-governmental organisation (NGO) that was acting in the case (AIDA – the Interamerican Association for Environmental Defense) and the lawyer representing the Consejo Territorial de Cabildos Indígenas de la Sierra Nevada de Santa Marta.

As I read the legal challenge in the case, I could not help thinking that there was something fundamentally perverse about this. The challenge had been promoted by developers and private interests and targeted the legality of Decree 1500, which protected the Black Line. It asserted that the protected territory was 'too wide' and that, when granted by a presidential decree, 'the Arhuaco, Kogi, Wiwa and Kankuamo' – the very groups benefiting from this recognition – had not been 'consulted'.

I dived into the documentation and did what barristers do: tested the logic of the argumentation and scrutinised the supporting material item by item. I found the basis of the challenge to be absurd. I saw that Decree 1500 had been adopted in compliance with a Constitutional Court Judgment,[22] whose direct antecedents were a series of cases brought before the court by the Arhuaco, Kogi, Wiwa and Kankuamo, in which their right to consultation had been ignored in the context of projects that negatively affected their ancestral territories. In addition, the Colombian Constitutional Court had acknowledged[23] that, during the armed conflict in Colombia, those indigenous peoples had been in danger of cultural or physical extermination, with forced displacement. In this context, their survival as a collective group was strongly linked to access to their ancestral territories, part of their identity and way of life.

Decree 1500 implemented what the Constitutional Court had held. In essence, it reversed a historical injustice. I decided to draft my intervention on the right to consultation coming from that angle. I focused on the important connection between Decree 1500 and previous Constitutional Court decisions ordering the executive to take certain measures in favour of the Arhuaco, Kogi, Wiwa and Kankuamo. I must have read thousands of pages in the preparation of my submissions, trying to extract what mattered from information that was not going to determine the case. To my mind the challenge was an abuse of process, because there had been a long consultation behind the Decree 1500 of nearly eight years duration, overseen by the Constitutional Court. Therefore, the conclusion that there had been due process was inescapable.

If the Decree were to be declared null, then effectively it would result in re-establishing the situation in favour of a third party: the private company with an interest in carrying out development projects in the territory.[24] The original case that exposed the violation of the right of the indigenous peoples of Sierra Nevada to be consulted referred to the project Port Brisa S.A. (a coal terminal). The Hukulwa hill, sacred to the Kogis, Arhuacos, Wiwas and Kankuamos, which stood guard of the ocean, had been demolished to build this coal terminal. The indigenous peoples warned them that this destruction of Hukulwa would upset the order of things, and that this would cause fierce storms and landslides. And it did. There is a scene in the Alan Ereira documentary that shows a group of Kogi Mamos confronting a crew building the port. The Kogi Mamos are barefoot, and they do not speak Spanish. They wished they spoke Spanish, they say, so as to be able to explain to the 'younger brother' the extent of their destruction.

When I saw this, I thought of my grandmother, a woman who never wanted to learn the language of the coloniser, Spanish. I admired her rebellious dismissal of a culture that had come to oppress the Andean world, that had come to oppress her. For my part, I, who grew up in Lima, never learned Quechua, so I never properly spoke to her. I remember you, grandmother, as the Kogis' image of Aluna, the essence within Nature, the Mother who conceived the world in darkness, spinning in silence, the thread of life.

Today, in the language of the coloniser, a language I came to master, I plead in courts of law for those whose side of the story (history) ought to be heard, as in this case. I argue from the viewpoint of the Arhuacos, Kogis, Wiwas and Kankuamos.

Decree 1500 was part of the implementation of a judgment that had acknowledged the right of these indigenous peoples to have their territory protected. So even if we were to accept (for the sake of *arguendo*) that the claimant in this new challenge against Decree 1500, an individual with no direct interest in the territory, could request its nullity, it could only be an abuse of right to promote the nullity of a decree that had been issued as a measure of compliance with a Constitutional Court judgment.

The claimant argued that Decree 1500 was issued 'irregularly' or based on 'a false reasoning'. But how can carrying out an order of the Constitutional Court lack lawful basis? The reasoning for the measure was found in the judgment. This request for nullity wanted to, indirectly, reopen and leave with no effect a judgment of the Constitutional Court. To me, it was an abuse of a procedural right.

In addition, international law also protected the space within the Black Line. When Colombia issued Decree Law 1500, it did so observing its own international obligations under a number of treaties and well-established principles. This included those principles reflected in the Rio Declaration, which states: 'Indigenous people and their communities and other local communities have a vital role in environmental management and development because of their knowledge and traditional practices. States should recognise and duly support their identity, culture and interests',[25] and also that: 'The environment and natural resources of people under oppression, domination and occupation shall be protected.'[26] Moreover, Colombia is a party to the 2015 Paris Agreement, which reinforces the importance of conserving and enhancing carbon sinks and reservoirs, including forests.[27] Preserving these forests and the natural world is therefore crucial not just for Colombia but also for the world.

For a week over Easter 2020 I was engrossed in the file and prepared submissions for a legal intervention as a non-party, or *amicus*, in the case. Such legal interventions are common in public interest cases, where the outcome is likely to have an impact beyond the parties. As I was studying the case, I wondered whether the indigenous peoples reaching out for help across miles of land and sea were the manifestation of the Mamos' (the enlightened spiritual leaders) intention to find support for the protection of their ancestral land.

The indigenous peoples of the Sierra Nevada believe that their mission is to be guardians of Nature.

The ancient guardians of the natural world

A press conference was called by the indigenous peoples, announcing the legal fight to preserve the ancestral land of the Sierra Nevada. I was invited to attend, marking the filing of the forty-two-page *amicus* I had prepared.[28]

'We are descendants of the great Tayrona nation,' said a Cabildo, a delegate for the indigenous groups. He was dressed in their traditional white clothes of maguey fibre. Surrounding him stood Mamos – spiritual leaders – wearing woven conical hats, worn in reverence to the snow-capped peaks of the sacred Sierra Nevada mountains. 'We [the Arhuacos, Wiwas, Kogis, Kankuamos] have a mandate which is in our Law of Origin: to care for and protect the elements that are the foundation of life in the universe: earth, water, air and fire,' stated the Cabildo.

Ten *amicus curiae* briefs – that is, legal interventions by legal experts, non-parties in the case (including the one I prepared) – were filed before the Colombian court on 15 July 2020, supporting their position. The indigenous peoples stated that their ancestors left them a knowledge system (pre-dating the Spanish conquest) and tools to maintain the harmony and equilibrium in the planet, an equilibrium that they believe is currently being lost.

Their efforts have so far preserved this irreplaceable area for the world. Among the many endangered treasures in the Sierra Nevada rainforest, several young scientists supported by the Conservation Leadership Programme rediscovered the Carrikeri Harlequin Frog (*Atelopus carrikei*) in a remote mountainous region, after fourteen years without sight of it. Amphibians are considered guardians of the planet, bioindicators of climate change, because the presence of frogs is linked to the purity of the water sources. The indigenous peoples of Sierra Nevada de Santa Marta, for their part, consider frogs to be a symbol of fertility and ecosystem health.

This is the world threatened today by land developers, mega-projects and the extractive industries behind the challenge to Decree 1500.

Like the amphibians who are believed to be guardians of the water, the indigenous peoples, the Arhuaco, Kogi, Wiwa and Kankuamo, are protectors of the future of the world. Their legal fight to preserve one of the world's most irreplaceable places matters to us and to future generations.

Reeds

Many months passed and there was no news on the case. We were waiting for a decision by the court on whether Decree 1500 should remain standing or whether it was to be nullified. The good news was that, although the claimant had sought a temporal suspension of the decree, this had not succeeded.

The process in Colombia otherwise was slow, moving at a snail's pace. Of course, since Decree 1500 continued to be in force, this did not affect the rights of the indigenous peoples of Sierra Nevada. However, the Arhuacos, Kogis, Wiwas and Kankuamos were entitled to legal certainty. The legal challenge (even as a process) purported to deprive them of this. As in the case of the Ch'orti's, the law protects the Arhuacos, Kogis, Wiwas and Kankuamos' position. Yet the Kafkaesque legal challenge to this protection demanded an incredible amount of energy to defend their territory, and could be replicated in further, equally burdensome legal challenges. I believed it was important to devise a way of acknowledging a supranational protection of that territory, permanently.

On a Friday evening in 2022, I had a video call with Yenny, a young lawyer from AIDA, the NGO acting in the case. I respected the work that she and several young Colombian lawyers and geographers were doing to protect the last wild areas in Colombia. I also discovered an invisible connection with Yenny – we had a surname in common. Not the surname that I carried, but my grandmother's surname, the surname of my ancestors: Junco, which translates as 'reed'. A reed is defined as 'a wild herbaceous plant with many straight, long, flexible and cylindrical stems, leaves reduced to a long and thin tongue, flowers grouped in almost terminal heads and fruit in a capsule'. We were wild herbaceous plants in another life, I said. She giggled.

'Have you considered pushing for UNESCO intangible status for La Línea Negra?' I asked.

I had advised AIDA on that before, in 2020, just as we had filed the *amicus*, including explaining the route and the timeline in which to achieve it. Only the Colombian state itself could put forward a nomination. But my suggestion was to elevate such a need to the relevant Colombian authorities. While any nomination would take a bit of time to progress, it was, in my view, the only way to stop the constant threats against the integrity of La Línea Negra. AIDA received this suggestion with interest. So had ABColombia, back in 2020. By 2022, I was still unaware if this advice had precipitated any action.

So, in my conversation with Yenny, I went over the mechanics once more. However, a synchronicity of some sort was already in place. My earlier advice had found its way to the Mamos. Things had already been in motion. A nomination by the Colombian government for the Sierra Nevada to be declared Intangible Cultural Heritage of Humanity had been made early in 2022.

Shortly after my conversation with Yenny, at the UNESCO 17th session of the Intergovernmental Committee for Safeguarding of the Intangible Cultural Heritage in Rabat in November 2022, the ancestral knowledge of the four indigenous peoples of the Sierra Nevada de Santa Marta (tied to the seas, rivers, stones, mountains and snow-capped peaks of the Sierra Nevada, known as La Línea Negra) entered the UNESCO List of Intangible Cultural Heritage of Humanity. Imagine visualising something, and then this becomes reality. That was the feeling I had. I was elated.

I was due to attend the ceremony in Rabat, Morocco, but unavoidable work commitments kept me in London in the end. So I watched the ceremony online. As the proposal of Colombia was announced as adopted by the Chairman of the states comprising UNESCO, a woman from the Arhuaco indigenous group spoke first. She was surrounded by other members of the indigenous peoples of the Sierra Nevada, all dressed in white, their traditional clothing. The men were carrying their *poporos* (a sacred gourd containing burnt seashells) and hand-woven *zijew* (small shoulder bags), where they traditionally keep coca leaves. 'Since the light of the world was created, indigenous peoples have had the mission of caring for the Earth [. . .] Our traditional knowledge is as valid and current as scientific knowledge,' the Arhuaco woman said in

Spanish, after some greetings in her native language. Another representative of the peoples of the Sierra Nevada stated: 'We have the legacy of caring for the heart of the world and life in the Universe, our principle of unity with Nature [. . .] Our responsibility is to transmit this knowledge to the new generations.' It was momentous and I felt proud of the important steps Colombia was taking to protect the Earth.

In our modern way of thinking, we assume that each generation knows more than the one before. In the conception of the peoples of Sierra Nevada, however, each generation knows less than the one before. We learn from our elders.

A new administration had taken office in Colombia, Gustavo Petro's, in August 2022. At the UNESCO ceremony, a member of the Colombian government ratified the commitment of the new administration in protecting the natural world as part of Colombia's fight against climate change.

The step taken in granting international intangible status to the knowledge and territory of the Arhuacos, Kogis, Wiwas and Kankuamo was part of that commitment. In fact, Gustavo Petro had gone to Sierra Nevada to meet its indigenous peoples, right after winning the presidential elections. There, he had received a blessing from the Mamos.[29]

While the new intangible status did not automatically stop the legal challenge against Decree 1500 (which is still pending), it has reinforced its legality under Colombian law, as this piece of legislation is in agreement with international commitments binding on Colombia. In Colombia, international law is directly applied by the courts. To put it another way, the possibility that the challenge to Decree 1500 might be successful is now very remote.

Colombia also took the step (together with Chile) of requesting an Advisory Opinion on the obligations of state members of the

Organisation of American States, in relation to Climate Emergency and Human Rights, before the Inter-American Court of Human Rights.

The Inter-American Court is one of three international courts that have been asked to clarify, for the first time, the legal obligations of states in relation to climate change and its impacts on present and future generations. It is a path that my work in the Torres Strait Islanders case, and the legal thinking behind it, set off. Climate change is being considered as a human rights issue, capable of being addressed as a matter affecting rights and triggering obligations and no longer as something restricted to the realm of mere discretionary policies by states.

The indigenous peoples of Sierra Nevada filed a writ before the Inter-American Court, so that their cosmovision would be taken into account in the Court's analysis of the state obligations in relation to climate change, in particular protecting all life in the natural world.[30] The participation of indigenous peoples in their own right in such international proceedings is unprecedented in any other similar proceedings in the world. For this reason, the Advisory Opinion of the Inter-American Court of Human Rights is likely to reflect a plural vision of the law, and not just a Western approach.

On 23 April 2024, the first day of hearings in the historic proceedings on the *Advisory Opinion on the Climate Emergency and Human Rights*, I found myself wigged and robed before a panel of six judges of the Inter-American Court of Human Rights, in Bridgetown, Barbados, opening my oral submissions, acting for a party focusing on the situation of small island Caribbean states:

> Madame President, members of the court, it is an honour to appear before you on behalf the Institute of Small and Micro

> States', I said. 'The UN Secretary-General, Mr Gueterres, has acknowledged that "[t]he Caribbean is ground zero for the global climate emergency". I respectfully submit that the meaning of "a real and imminent risk" to the right to life in that context is the one formulated by the Dutch Supreme Court in *Urgenda*. The term "immediate" does not refer to imminence in the sense that the risk must materialise within a short period of time like the Human Rights Committee wrongly concluded in *Teitiota* and the *Torres Strait Islanders* case. But rather, that the risk in question is directly threatening the persons involved. This is what is happening with the population of Small Island States like the Caribbean countries which contribute to only 1 per cent of global greenhouse gas emissions yet stand to be disproportionally affected. I invite this court to adopt the *Urgenda* test and clarify the criteria in which the right to life is breached in the context of climate change.

This emphasis was premonitory. A few months after the hearing Hurricane Beryl, a deadly and destructive tropical cyclone, seriously impacted parts of the Caribbean. The court had intentionally held the hearings in the Caribbean and Brazil so as to hear from those most affected and at risk from climate change: indigenous populations, coastal communities, island states, states affected by deforestation and loss of biodiversity. A total of 265 written submissions had been filed before the court. More than 150 oral interventions were received from states, international and national organisations and other parties at Bridgetown and Brazil. My submissions had focused, among others, on the right to life and the meaning of the term 'imminence' in the context of climate change; the very notions that had been left in an unsatisfactory state in the Torres Strait Islanders case.

My experience with the fight of the Arhuacos, Kogis, Wiwas and Kankuamo, and other cases, has taught me about resilience and that the law advances in an incremental way. Each step to influence the law in the right direction is important and worthwhile pursuing.

Of course, time is pressing. In their brief before the Inter-American Court, the Wiwas and Kankuamos denounced the way that time, seasons, were now out of control (*la descontrolación del tiempo*) and birds that announced the time to cultivate coffee or avocado are no longer to be seen. They have gone higher up, to colder weather. I was reminded of what a Mamo had stated in an interview: 'The Younger Brother [as they call people in the West] is damaging the world. He is on the path to destruction. He must understand and change his ways, or the world will die.'[31]

EIGHT

The Masewal and the Constitutionality of Federal Mining Law in Mexico

Who owns the subsoil, the ground beneath your feet?

I was compelled to contemplate this question more deeply when I first met Ximena, a lawyer working for the Mexican Centre for Environmental Law (CEMDA), and heard about her case representing the Masewal people.

The Masewal, based in the Sierra Norte in the state of Puebla, believe that the soil that we walk on and the Earth itself is 'alive' (*nemilis*).[1] But in Mexico, the Federal Mining Law assumes that the State owns the subsoil and it makes no reference to the right of consultation of indigenous peoples in the process of granting of mining concessions.

Now, however, the Supreme Court of Justice of the Nation (SCJN) of Mexico was examining a challenge brought by the Masewal people of Cuetzalan, who had filed a constitutional action against Mexico's Federal Mining Law. A central question was whether the law was unconstitutional, because it should have followed a consultation process with indigenous peoples prior to its being issued – and how it fitted within the constitutional and international law framework. The Masewal requested that the SCJN declare the law unconstitutional because it violated indigenous peoples' fundamental rights.

Ximena was advising the Masewal people in this case and we met online in July 2020 to discuss an *amicus curiae* intervention in the case: an independent, non-party brief to assist a court, by offering expertise in legal questions crucial to the determination of the case. It is particularly allowed in Latin American systems when there are public interests involved in the case. Ximena invited me to file one, for which she would facilitate all documentation.

There was something very refreshing about Ximena. She belonged to a new generation of Latin American women lawyers, with Master's degrees in environmental law, working on some of the most crucial issues in Latin America. The topics on which she was working were pressing, yet she didn't appear stressed, more as if she had all the time in the world. She lacked the air of self importance lawyers sometimes assume, and her Twitter account, 'out-of-context Ximena', was humorous and fun. In a world overburdened by self-promotion, this felt refreshing to me.

As I explored the case further, I understood that there was a need to understand the beliefs of the Masewal, to fully comprehend the impact that destruction of the soil had when you believed that the Earth was alive.

Who owns the soil and land?

The Masewal (or Nahua, as they are also known) live in the highlands of the State of Puebla, in the mountainous region of Mexico called La Sierra Madre Oriental. *Masewal* is locally translated as 'people from the earth' (*gente de la tierra*).[2]

The Masewal think of the Earth as a living entity. In the Masewal's belief system, 'the stones are [the Earth's] bones and that is what sustains them [. . .] That is why we say that stones are important.'[3] They speak to the Earth.[4] They treat her as delicate and avoid chemicals in agriculture because they believe 'they poison her'.

'Being alive, [the Earth] is also sensitive, she listens when they speak to her through the caves (*ostokj*) and she also breathes through there. For the *maselwalmeh*, caves are places that we fear and respect because they are the entrance to the underworld (*Talokan*).'[5]

Further to this, they believe that the soil (which is what we walk on) is alive, and that humans form part of the soil. 'Matter is not destroyed, it is only transformed. We are what we eat.'[6]

The Masewal have a *Códice*: a code of life. At its heart it is *yeknemilis* (*La Vida Buena*) – or 'good living' – which summarises a way of life in harmony with Nature. From its cosmovision, the Earth is not 'owned' by anyone. It is a living entity and sacred.

In my research I came across the work of an academic called Alessandro Questa, who has written ethnographic studies about the Masewal people. He states:

> Different Masewal interlocutors have told me that they see themselves either as 'in charge' of the surface (not its depths) of the earth or *tlaltikpak*, or as its 'tenants'. Masewal farmers talk to their 'partners' and their *compadres*, the invisible subterranean beings who grow and harvest the maize together with them. The true 'owners' or *itekomej* and 'hosts' or *chanekej* or *chanchiwanej* are the non-human ancestral beings who give life to the land around them and who are the force behind most of its continuities and transformations.[7]

How is one to resolve these two contradictory approaches to the Earth? The one reflected in the Federal Mining Law and the one of the Masewal? Happily, international law of indigenous peoples' rights provides a framework to resolve this conflict, and acknowledges indigenous peoples' right to be consulted over projects that may impact their ancestral land.

I started drafting my *amicus*. I confess this was a technical endeavour, an analysis of provisions, hierarchies and interpretative techniques that may sound dry to most people. But while I was working my way through these technical details, I was intrigued by the Masewal and their systems of belief. It intrigued me because it invited me to think differently. Why should we assume that the only way to look at the Earth, the 'serious' way, is the Western approach, and that the only relation to the Earth is one of 'exploitation'? By the time I was engaged in this case, I had already witnessed through my work a massive negative impact on, and the destruction of, the natural world. Why do we treat the Earth as something inert, devoid of life?

Despite its initial strangeness, the Masewal's approach seemed sensible to me. Their acknowledgement that there are forces at play that give life to the land around them, and that are behind most of its continuities and transformations, appears to me to be consistent with the scientific acceptance that there exist laws and principles ensuring the proper functioning of the natural world. Shouldn't we respect these laws in any interaction with the Earth?

The Masewal's approach is in addition one that only considers our human, individual relation to the land as being temporal, not as 'owners' of the Earth. Wouldn't it sound ludicrous to us if we heard someone purport to 'own' the Moon?

I set out to learn more about the Masewal.

Water

Extractivism is an economic model consisting of the removal of 'natural resources' from the land or from underground in order to put these resources up for sale as commodities in the global market. One of the great concerns of the Masewal in the context of extractivism in their ancestral lands is its impact on water. Like the Earth, water is 'alive' for them.

As observed by Questa, 'according to [the Masewal's] native ecology, the clouds use the land to move between the forests and ascend to the sky only to return to "sleep" on the hills each night, forming the dense fog, characteristic of the region'.[8] Questa observes further that, in the Masewal's cosmovision, if the water emerges from the subsoil, it does so 'with the intention of sprouting for the people and thus returning to the clouds and returning again in the form of rain and mist'.[9] He points out that the Masewal's water cycle (similar to the scientific one) considers, however, that the water is alive and that it responds to the designs of its spiritual patrons, who move or hide it. [10] The winds, for their part, are born in the distant mountains and, when they collide with others, they bring with them clouds and rain, 'sent by the enormous will of invisible aerial entities'.[11]

In the Masewal account of the Creation, *Sipaketle*, or the water deity (*la dueña del agua*), 'throws herself with such force against a non-existent hill that it makes a large hole in the ground; her body, transformed into salt water, generates the sea, and her hair becomes underground rivers that will gush out at various points on earth, producing the innumerable springs that provide drinking water to human beings'.[12]

The *Sipaketle* made it possible for humankind to have the Seed (the fertilised ripened ovule of a flowering plant containing an embryo and capable of germination to produce a new plant). In the Masewal's vision of the Creation, the Seed is sacred. Before planting, it is blessed. The celebration of the Blessing of the Seed and the Water Festival are related festivities.[13] In both celebrations the *Sipaketle*, or the owner of the water, appears as the central figure. Both celebrations are carried out inside the caves, which are assumed to be, or were, channels formed by water, and which mostly communicate with underground rivers. The Water Festival is held on 3 May in the caves, spaces known as meeting places with the *Sipaketle*, and marks the beginning of the rainy season.[14]

I am not surprised about this cosmovision where Water, Wind or Earth are regarded as living entities. The American biologist E. O. Wilson once referred to Nature in its wild state as 'self-willed',[15] and warned us of the 'erroneous belief that the living environment is less important to humanity than the non-living environment'.[16] The question is: when did we lose this respect for the Earth and for Water that is so present in ancient cultures?

As I conducted my research on the case, learning of the Masewal's sacred connection to water, contrasting images came to mind. Images of miles and miles of tonnes of plastic waste dumped in the Malaysian jungle, shipped out from England: first-world waste.[17] Tonnes of unused, unsold clothes made with non-biodegradable chemicals, shipped from Europe and the US, and dumped in the Atacama Desert.[18] Fast fashion tanneries polluting the water supplies in India with toxic chromium, making their way into cow's milk and agricultural products.[19] We treat Nature as a dumping site and poison ourselves in the process.

When did we stop looking at a seed as something sacred?

The birth of mountains

I am impressed by the correlations in indigenous views of the world. For my ancestors, mountains are sacred and the home of the Apus (the spirit of the mountains). To the Masewal, mountains also occupy a prominent place in their ontology. Each mountain is 'a sentient entity', a reservoir of life, 'a container for other bodies and laboratory for novel forms [. . .] Every technological achievement, every tool [. . .] is originally taken from a mountain.'[20]

For the Masewal the world was fragmented long ago by the Awewe.[21] In Masewal cosmovision, the Awewe 'is [. . .] the older brother of [. . .] the sun, whom [the Awewe] envied, wishing to be praised in his place'.[22] To the Masewal the mountains are remnants of pillars that connected the sky and the Earth:

> The spiteful Awewe crashed into each of the many pillars that connected heaven and earth, fragmenting them. Until then, people could climb the pillars when they felt old and sick, and then climb down again as children. In other words, people did not know death. The Awewe was finally banished, thanks to a trick that made him crash into pillars made of clouds and ended up at the bottom of the sea, from where he still causes storms and hurricanes with his thunderous roars. The rest of the old gods remained in the sky, and people, of course, on the surface of the earth. The destruction of the pillars implied the partial fracture of a combined world and the appearance of two incomplete but interdependent worlds, in which ethereal spirits and humans continue to interact with each other. Indeed, the collapse of the pillars did not totally split heaven from earth as some beings can still cross from one side to the other.[23]

This way of looking at mountains fascinates me. As someone who grew up by the sea, I only discovered mountains, properly speaking, with the European Alps. The top of a mountain is indeed a place where a special energy seems to have existed since time immemorial. It was Klemens, my partner, a keen climber, who took me walking around the many mountains of the Austrian Alps, where reaching the top feels like reaching some sort of heaven.

This experience with mountains took me some years back, to my reading of the poetry of ancient Chinese hermits, who chose mountains as their abode to retreat from the world and seek enlightenment, hermits such as the Chinese Buddhist and Taoist monk Hanshan, or Cold Mountain, who lived during the Tang dynasty, and the

Chinese Zen monk Stonehouse (1272–1352). Taoists mountain ascetics lived in cliffs and caves, a secluded, simple life, seeking to understand the Tao, the primordial laws or natural way of the universe. The ancient Chinese regarded these sages as able 'to call up the wind' and 'move between heaven and earth', or as 'immortals'. Hermits in the mountains 'could talk to heaven'.[24]

Returning to the Masewal example, the mountains, seen as the gigantic remnants of the original pillars – as broken and interrupted structures that once reached Heaven – are perceived as a link between worlds. This is why, for the Masewal, the tops of the mountains are still related to 'the spirits of the sky.'[25] Those studying their practices note that 'fortune tellers and healers climb, above all, the highest peaks such as Kosoltepetl, Tonaltepetl, or Chignamasatl, to be able to see from afar and from above, leave offerings to the "angels", *ejekamej* or spirits of the wind, and ask them to carry messages, to bring good weather or to maintain the sustenance of life.'[26] In the Bolivian Andes, a similar practice takes place. In the 2022 Bolivian film *Utama*, when climate change, with its lack of water, makes life unbearable for an old couple who live on the edge of Salar de Uyuni, the world's largest salt plain dry region, Virginio, a character in the film, goes up to the mountains, as a last resort, to perform traditional rituals to ask the Apus, the living spirit of the mountain, for much needed rain.

It appears wondrous to me that mountain landscapes shaped into forms of enlightenment were also reflected in the 'rivers-and-mountains' (*shan-shui*) poetry tradition in China, which began in the fifth century AD. This tradition is the earliest human literary engagement with the idea of wilderness.[27] It is a poetry rooted in Taoist and Zen thought, where humans are part of the flow of the natural world and its processes. As noted by Hinton, '[T]he Chinese wilderness is nothing less than a dynamic cosmology in which humans participate in the most fundamental ways.'[28] Just as the Masewal participate in the natural world in which they are immersed.

I spent many years, well before my work at the Bar, reading the poetry of Li Po, Tu Fu, Wang Wei, Wei Ying-Wu and others. It connected me with another way of thinking about the world. I often experienced the impossibility of communicating what occupied my mind, what really mattered to me in those days to anyone 'in the dusty world', randomly striking conversation with me. Did these readings enabled me to carry out better my environmental work years later? To understand people like the Masewal?

From the Tang dynasty poets, a favourite of mine is Li Po, who wrote:

You ask why I make my home in the mountain forest,
and I smile, and am silent,
and even my soul remains quiet:
it lives in the other world
which no one owns
The peach trees blossom,
the water flows.[29]

I find peculiar as well that, in the *Upanishads*, the ancient texts from India, the creation of human life is described as the creation of 'world-guardians'.[30] Much in the same way that the Masewal see themselves. The *Upanishads* also perceive at the core of human life 'an interrelationship between man and nature'.[31]

If humanity's elders were clear about all of this, and that the underlying force of all life was the natural way of the universe, shouldn't humanity's laws, which affect all that lives on Earth, be in tune with those eternal laws? This is an essential proposition put by Cormac Cullinan in *Wild Law*,[32] which strikes me as correct. He posits that human laws must be designed to promote behaviour that contributes to the health and integrity not only of human society, but also the wider ecological communities, and of Earth itself.[33] Put another way, human laws can only develop in harmony with the laws of Nature.

My burning question is therefore: when did humankind lose its way? And can we retrace our path?

A victory for the Masewal

Ximena forwarded all the documentation that was part of the file of the case to me and I started my work of going through domestic pieces of legislation, interpreting constitutional provisions, bringing international law into the equation, and drafting my brief. A barrister never works at a leisurely pace (at least not most of the ones I know). This exercise of digesting huge amounts of information and shaping legal argument usually happens under tight constraints. If you find law attractive as a career, ask yourself whether you like to read. Reading, broad reading, will prepare you for such a career. You have to comprehend things and penetrate matters relatively fast. You have to be able to find what matters from among thousands of papers and distil in one question what the central issue is in a dispute if you are to be effective.

I eventually filed my *amicus curiae*, or 'friend of the court' brief, on 1 September 2020. The Court, having heard the parties already, admitted it. My writ offered the Court impartial, specialist knowledge to assist them in resolving the dispute.

On 13 January 2021, the Supreme Court issued a judgment. It reaffirmed that, in accordance with the Constitution of Mexico and the Convention 169 on Indigenous and Tribal Peoples of the International Labour Organization, all the authorities of Mexico, within the scope of their respective competence, 'are obliged to consult

[. . .] indigenous peoples before adopting any action or measure likely to affect their rights and interests'. It also provided that the Mining Law ought to be read in accordance with such an obligation.[34] When I was in touch with Ximena later, she told me: 'We are prepared to fight this one, all the way to international courts if necessary so that this is applied.'

But it was not necessary. Following all this, on 16 March 2022, the Masewal achieved a major victory against mining concessions in their territory. The Administrative Court of the Sixth Circuit delivered a judgment in which it agreed with the Masewal people in their lawsuit against three mining concessions in the state of Puebla. The Court held that the issuance of these concessions violated their rights to consultation and prior, free and informed consent, and for the same reason they ordered that they should be left with no effect and be cancelled.[35]

In its press release, CEMDA stated:

> For the Masewal people, the land is a sacred site because there is the Talokan, which is the place where the lords of life dwell and where seeds, plants, animals, water and fire are protected, and the Tlaltipak, *la tierrita* (the little land), which encompasses everything that lives on the ground. If the mining project is carried out in its territory (altepet), the Talokan will cease to exist and what is in it, such as the cornfield (milah), the intervened hill or forest (kwohtah), the hill in which it is produced (kwohtakiloyan), the pasture (ixtawat), the shaded coffee plantation (kaffentah), and sacred sites, such as caves and springs (ameyalmeh and apameh) will gradually disappear until everything essential for good living (yeknemilis) is finished.[36]

To the Masewal what is threatened is life itself.

Dancing for the climate

According to one of the Masewal's myths, the world is 'already broken'.[37] Questa explains that, to the Masewal, 'the world already broken, is somehow breaking apart yet again, this time under the pressures of a kind of immoral Anthropocene.'[38] The Masewal experience the current environmental and climatic crisis as 'a moral crisis', and as the result of humanity's 'breaking ancient accords' and 'disrespectful behaviour'.[39]

There is no word for 'poverty' in Náhualt.[40] 'Poverty is locally understood not as a lack of money but as an absence of life'.[41] In their system of values, therefore, life is at the first priority.

But isn't it the same in the system of values reflected in all legal systems of the world?

The Masewal see the need to fix a damaged world, including changing people's disrespectful behaviour.[42] They ascend the mountains and connect with its spirits. In the context of climate change and mining exploitation threats, the Masewal dance (*Tipekayomej*, *Wewentiyo*) 'to reconnect with their spiritual landscape and visualize alternative futures'.[43] These dances are ways of thinking about the world.[44] They dance as a way 'to 'remember' ancient forms of negotiation with the *itekomej*, or 'owners' of the land.'[45] These ancient forms of negotiation are not just specific to the Masewal. They also exist in the Andean and Amazonian tribal worlds. The indigenous peoples there don't just 'take' from the Earth. They also give to the Earth (*pagos* or offerings). Spiritual leaders in communities, known as *Layqas* in the Quechua world, or shamans in the Amazonian tribal world, 'negotiate', through their rituals and practices, what humans may take, so that a proper balance is maintained between the needs of the human community and the Earth. Curiously, I sometimes feel that I am a sort of legal *Layqa* or shaman who, through my work, 'negotiates' what humans may take, ensuring that the limits are respected.

Lastly, there is something about the Masewal's dances worth stressing. When visualising alternative futures, these dances reflect a non-anthropocentric and multispecies conception of society.[46] They are, of course, not the only ones in the Americas that have such a non-anthropocentric vision of the world. In *El Pez de Oro* (The gold fish),[47] a work of great literary rarity published in 1957 by Arequipa-born Gamaliel Churata, the Andean world is vindicated as non-anthropocentric and multispecies, where the root of humanity is acknowledged to be deeply connected to the animal world, and other forms of life (originating in the water), and where the human and non-human (Earth, vicuñas, lakes, mountains, sirens, frogs, ice, pumas, stray dogs), co-exist as part of an ever flowing fabric of life. In this Andean cosmovision reflected in *El Pez de Oro*, in Churata's own words, humans 'live in perpetual dialogue with nature',[48] and the Andean man 'has had to overturn and confine himself within and in the very bowels of the soil [. . .] has extracted his conception of the world and of himself from the land with which he has interacted in century-long dialogue'.[49]

Churata remained ignored and unappreciated by the literary establishment for not reflecting the traditional Western canons during his lifetime. He died forgotten – in Lima – where he had returned from exile. Declared Cultural Heritage of the Nation in Peru nearly seventy years after publication, in 2024, *El Pez de Oro*, his work – perhaps one of the most important literary works to emerge from Latin America – provides a window to another way of perceiving the world.

NINE

The Rights of Nature Case: Los Cedros Cloud Forest

Can a planet have legal rights? Could it be defended in a court of law? And who or what are we referring to when we refer to the right to *life*?

By 2020, these were some of the essential questions that the work I was doing pushed me to ask myself. A new case concerning a mining concession had landed on my desk. It was unusual because, unlike most of the cases in which I had been involved before, there were no people in the area where the mining activity was going to take place. The case was about a cloud forest called Los Cedros, and it was about defending an ecosystem in its own right.

It is not every day that an English barrister goes to court to argue a case for the rights of Nature itself. But in mid-August 2020, I found myself acting on this case that focused precisely on that. I had been invited to act as *amicus curiae* before the Constitutional Court of Ecuador, immersing me in the natural world at risk of extinction in a cloud forest near the equator.

The case concerned a mining concession to explore gold and other minerals and a cloud forest. But unlike some of the cases that arise from Latin America, there were no indigenous peoples inhabiting the affected forest. Therefore, the government had taken the view that there was no one to consult on the project, and, as a result, it had granted the concession. This made the entire case focus on whether the Rights of Nature would be violated by the destruction of this cloud forest with all its living species, both fauna and flora.

It was an unprecedented case and the experience of acting in it was illuminating.

Rights of Nature

While debates take place around the world about whether or not Nature should have rights, in Ecuador it already does.

In 2008, the Constitution of Ecuador adopted a unique approach that abandoned an anthropocentric approach and became the first country in the world to recognise the Rights of Nature. Nature is considered 'a living being'[1] with inherent constitutional rights.[2] The constitution provides that 'Nature, where life is reproduced and occurs, has the right to integral respect for its existence and for the maintenance and regeneration of its life cycles, structure, functions and evolutionary processes.'[3] Therefore, Ecuador could become the first country to protect large areas of biodiversity based upon this constitutional innovation, setting a precedent worldwide.

This legal provision reflects the cosmovision of indigenous peoples in the Americas. The preamble to the constitution makes it clear that this position has its roots in an ancient tradition of its own, where the original peoples conceived themselves as part of Nature, of the Pacha Mama, which was considered 'vital for our existence'. It expresses the sovereign decision of the Republic of Ecuador to build 'a new form of citizen coexistence, in diversity and harmony with nature, to achieve good living, *sumak kawsay*'. These principles run through the entire constitution, which is guided by comprehensive respect for the existence of Nature, its biodiversity, its maintenance and its regeneration. Therefore, development is not conceived in detriment of these principles, but in line with the concept of living in harmony with the natural world.

In a context in which humanity faces an existential challenge with global warming and climate change, the model followed by the Constitution of Ecuador offers an important perspective for the world.

In a ruling dated 18 May 2020, the Constitutional Court of Ecuador selected the case of *Los Cedros* to be heard, becoming the first world case in which a court addressed, in a central manner, the scope of the Rights of Nature. Effectively, it became a landmark ruling on what is a growing body of law called Earth jurisprudence.

Irreplaceable

In *Irreplaceable*, Julian Hoffman explores the notion of *topophilia*, or 'love for place [. . .] our instinctive desire to forge attachments to landscapes that impart personal meaning, value and identity as they intertwine with our lives and communities'.[4] For example, we love a 'place that has been experienced deeply in our lives'.[5] A place where we have developed 'indelible ties with'.[6] We also develop ties with specific plants, smells, topography.

In 2003 I saw lemon trees again, in Mexico City, after a long time of not having seen them, as I'd been living in England. I felt joy, and could not stop taking pictures of them. I felt like clapping. The Mexicans who were in the surroundings looked at each other, wondering, perhaps, whether there was something precious hidden in the tree. Lemons were too ordinary to them to spark such a reaction. Similarly, when I visited the Exotic Garden in Wisley, Surrey, I saw tall banana trees again after not seeing any for a very long time and felt suddenly happy. I sat among them and did not want to leave. I recalled the banana tree and small garden in one of the houses where I lived in Peru. Back then I had not thought much of the tree. But the sight of the leaves sent me straight there, to the tree my father had planted with his own hands. I did not want to leave the garden.

But have you ever come to love a place you have never actually been to? As I immersed myself in the ecology of Los Cedros from across the Atlantic, I felt a similar feeling of connection. I got to know it and to value its place in the world. Even if I may never step foot in it, I could picture Los Cedros enveloped in a hazy mist, in a thick veil of clouds. I know the smell of a cloud forest. I have been in Monteverde in Costa Rica, where the vegetation is lush, the rainfall is heavy, and the condensation is persistent. You feel the moisture in the air when you breathe. You find yourself surrounded by epiphytes: lichens, orchids and bromeliads. There is a unique feeling, as if you are enveloped in a magic world.

As I drafted my brief from my London flat, on the hottest day in August, I marvelled at the multitudes of life existing in the forest at the heart of the dispute. I felt honoured to have been given this unusual task of being a barrister for the Earth, arguing for the right of existence of this irreplaceable living world.

Ecuador has some of the greatest biodiversity on the planet, despite being only a small fraction of the world's land surface.[7] The tropical Andes of Ecuador are, in fact, 'at the top of the world list of biodiversity hotspots in terms of vertebrate species, endemic vertebrates and endemic plants'.[8] This is partly due to its location; scientific literature notes that 'biodiversity increases towards the equator and decreases towards the poles'.[9] Thus, more or less half of all plant species are found in tropical forests, which represent only 7 per cent of the world's total land surface.[10]

Los Cedros, located in the tropical Andes of north-western Ecuador, has three rivers and nearly 12,000 acres of primeval forest. Primeval forests are the oldest forests on the planet that have not been transformed or altered by human activity, and Los Cedros is one of the few remaining. They perform a key role in the climatic balance of the world, not only important for the oxygen they provide us but also for their ability to absorb CO_2 (carbon dioxide) from the atmosphere, thus mitigating climate change. Plus, in no other terrestrial ecosystem are there so many different types of animals and plants as in primeval forests. They are estimated to be home to about two-thirds of the world's terrestrial species of plants and animals, many of them still unknown.

Los Cedros itself is thought to be home to 299 tree species per hectare, 400 species of orchids (only 187 identified so far) and many notable rainfrogs (almost all in danger of extinction). Los Cedros is also home to 317 species of birds, including the cloud forest Pygmy owl, and six wild cat species, including jaguars. It is the home of many endemic species, including the last critically endangered brown-headed spider monkeys in the world, as well as the endangered Andean spectacled bear, which, as it happens, inspired the beloved fictional character Paddington Bear.

The elevation of Los Cedros varies from 980 to 2,200 metres, which places it entirely in the zone of lower montane rainforest, also known as cloud forest.[11] The 'cloud forest' is characterised by a high concentration of surface fog, usually at the level of the canopy.[12] Scientists note, however, that Los Cedros is located in 'one of the most threatened eco-regions on the planet.'[13] The distribution of a species is limited to a small geographical area due to a combination of microclimatic or topographic barriers.[14] For example, scientific research notes that, in Ecuador, 27 per cent of the known plants are endemic, and many of the endemic species 'are known from only one or a few localities in a single province, and are thus not found anywhere else in the world'.[15]

Endemism rates are higher in the mountains than in the lowlands.[16] This observation is directly relevant to understanding the biodiversity of the Los Cedros Protected Forest and its vulnerability to any mining project. Los Cedros protects a unique subset of various taxonomic groups in this region of highly localised endemics.[17] This means that dramatic changes in species composition occur within them, on short spatial scales.[18] The affectation/degradation of a relatively small piece of land in the Cedros can extinguish species that do not exist in any other part of the world.

The diversity in Los Cedros – in addition to its being a primeval cloud forest – can also be attributed to the stability of its habitat. Both fossil and phylogenetic methods suggest that some lineages present in the forest have been around for 67 to 115 million years, indicating climatic stability and low extinction rates.[19] Los Cedros therefore acts as 'a museum of biodiversity accumulated over a long time in the lowlands and as a cradle of new adaptations and speciation spurred by the uplift of the Andes.'[20]

This is the world that Cornerstone, a Canadian company, was aiming to mine, with the authorisation of the Ecuadorian government. I am rather disturbed by humanity's capacity to destroy. Pulitzer Prize Awardee, biologist Edward Wilson warned: 'we're extinguishing Earth's biodiversity as though the species of the natural world are no better than weeds and kitchen vermin. Have we no shame?'[21] To him, 'the study of every kind of organism matters, everywhere in the world'.[22] I am starting to see it in all its beauty.

The biologist Elisa Levy Ortiz observes: 'Los Cedros also constitutes one of the last remnants of these mega-diverse forests. The forests of western Ecuador, including cloud forests, have disappeared. In the year 2000, it was estimated that more than 96 per cent of the primeval forest land in western Ecuador had been deforested [. . .] and much of this remaining 4 per cent has since been lost.'[23]

Los Cedros is not accessible by road, and for this reason it has, until now, been better protected and less scientifically explored than some other Protected Forests in Ecuador.[24]

Connectivity

As I research the case, particular species caught my attention. I am fascinated by frogs, especially rainfrogs, and I learn that the frogs of Los Cedros are 'remarkable, almost all endangered and only found in the local cloud forests'.[25] The *Pristimantis cedros* species, scientists note, 'is locally common at Los Cedros but has not been collected elsewhere'.[26] The *Pristimantis mutabilis* is 'only known from two streams, one of which is at Los Cedros'.[27] This unusual frog 'is capable of changing the texture of its skin', something never seen in frogs before.[28]

Los Cedros is also extraordinarily rich in plant species, with at least 299 species of trees per hectare (many of them endemic).[29] Associated with this forest are many fungi (some endangered) that are essential for the growth of the forest. A recently described plant, discovered at Los Cedros, *Cuatresia physalana*,[30] 'is related to tomatoes and potatoes, and therefore may contain valuable genetic materials valuable for agriculture. Furthermore, *Cuatresia* are known to contain antimalarial compounds.'[31] This genetic and species diversity is likely to be erased from the Earth if the project is to go forward.

I focus my attention on one particular species: the Andean or spectacled bear (*tremarctos ornatus*), also known across the world as Paddington Bear. The Andean bear lives along the Andes mountains, across Peru, Ecuador, Colombia, Bolivia and Venezuela. It is a species of great importance for the conservation of other plant species, since it is considered an umbrella species, because it eats and walks, propagating through the forest seeds of various plant species.[32] But its habitat is rapidly, dangerously shrinking.

Up to the point that the Los Cedros case started in a lower court in Imbabura, it was uncertain whether bears existed in the forest, and the full extent of what was at risk. Was the Andean bear at risk?

The Andean bear is generally elusive. It is nocturnal and primarily vegetarian, eating fruit, berries, cacti and honey.

An on-site inspection by a relevant authority from the municipality was carried out during the lower court proceedings to ascertain whether there had been any damage in Los Cedros as a consequence of the initial work the mining company had begun in its exploration. The inspection discovered serious damage by some initial exploration, in an intersection with the Los Cedros Protected Forest.[33]

It was then that the technician from the municipality carrying out the inspection saw it: footprints of the Andean or spectacled bear could be seen on the ground (the footprint of five bear's toes with strong, curved claws). He followed the footprints and observed Andean bear faeces within the trail travelled. This corroborated the footprint found.[34]

This – which must have felt like an exciting discovery for the scientific community – was the confirmation that the Andean bear (or spectacled bear) existed in Los Cedros.

The existence of this species of mammal, classified as endangered,[35] had not been taken into account by the Environmental Management Plan of the Río Magdalena Mining Concession (the official name of the mining concession) when it had granted the concession.[36]

The Andean bear is a species of mammal in the Ursidae family.[37] It is the only living species of its genus.[38] It appears as a vulnerable species on the International Union for Conservation of Nature (IUCN) Red List.[39] The destruction or 'fragmentation' of its habitat is one of the main reasons why the Andean bear is in danger of extinction.[40] As noted by Edward O. Wilson generally, 'every expansion of human activity reduces the population size of more and more species, raising their vulnerability and the rate of extinction accordingly'.[41] This was the first time that I became aware that, when a species loses part of its range, it is in greater danger of extinction[42] and that 'the chance that a population of organisms will become extinct in a given year increases as its living space is cut back'.[43]

For the Andean bear to have a chance of avoiding extinction, therefore, it was not enough to save Los Cedros. The forests around it, every bit of them, were equally important. The science clearly tells us that 'when a piece of primeval forest is set aside and the surrounding forest cleared, it becomes an island in an agricultural sea. Like wave-lapped Puerto Rico or Bali, it has lost most of its connections with other natural land habitats from where new immigrations can occur.'[44]

The isolation of habitats is a concern for biologists worldwide. Species need to be able to move from one area to another. The Andean bear, for example, can walk fifteen kilometres in a day. The preservation of biodiversity and fauna population as a whole depends on the ability to move.

I became increasingly aware that Los Cedros formed part of a larger ecosystem. It borders the Cotacachi-Cayapas Ecological Reserve, protected at national level in Ecuador.[45] Scientific studies note that Los Cedros is part of a buffer zone and a southern corridor for the still protected Cotacachi-Cayapas Ecological Reserve.[46] A buffer zone in this context is a connecting zone that links two or more protected areas. I read with apprehension a scientific study pointing out that Cotacachi is in danger of becoming an island surrounded by mines.[47] Was this the future of the Andean bear if we were not to act?

'Islands lose diversity and ecosystem services because of increased isolation and edge effects such as forest drying and increased predation,' note the scientific studies.[48] In essence, the permanent protection of Los Cedros and its linkage with Cotacachi-Cayapas was essential for the functioning of the western Andean corridors (the Andean Chocó, Andean Bear Corridor and the recently proposed Biosphere Reserve) due to the large differences in elevation.

Another reason for maintaining the corridors is the 'rapid and ongoing climate change'.[49] As the climate warms and dries, scientists note that connections between lower elevations and higher elevations become necessary for the migration of organisms that respond to it, as is already occurring in Ecuador.[50] Los Cedros is in that sense ideally located to form the connecting point of a southern corridor to Cotacachi.[51] This corridor includes preferred habitats for the most threatened species including the primates,[52] cats,[53] bears,[54] as well as frogs,[55] birds[56] and orchids.[57] As Edward O. Wilson says, 'life is an exceedingly improbable state, metastable, open to other systems, thus ephemeral – and worth any price to keep'.[58]

Los Cedros as a source and water recharge zone

I turn to water sources in Los Cedros and find that the forest 'protects the origins of three rivers' – the Manduriacu River, the Verde River and the Los Cedros River – in addition to encompassing the southern bank of the upper Magdalena Chico River.[59] Scientists indicate that 'these rivers supply freshwater to people and are the habitat of an amazing diversity of life.'[60] The *amicus* of the biologist Elisa Levy Ortiz, in the Protection Action before the Multi-competent Chamber of the Imbabura Provincial Court, indicated that these rivers 'supply water to the people below, and are the habitat of an incredible biodiversity of life themselves'.[61] Scientists recorded that 'in a three-night exploratory survey, almost forty species of caddisflies (*Trichoptera*) were collected, more than a third of which are probably new to science'.[62]

Elisa Levy Ortiz also indicated that the best way to measure water quality is with the use of aquatic macroinvertebrates, since they live in the water and affect physical and chemical aspects of water bodies, and that 'a recent study in Ecuador concluded that water quality is excellent in Andean and foothill streams only when the basins had an intact native vegetation cover of at least 70 per cent'.[63]

The importance of Los Cedros for the water cycle

As I made my way through the scientific material, I discovered that Los Cedros is part of the fundamental hydrological cycle in Ecuador and plays a key role in providing pure water and generating rain. Scientific evidence shows that cloud forests (including Los Cedros) are 'key to both maintaining the production of pure water as well as to capture and purify water, since they regulate the water flow at the landscape level and, likewise, it is the vegetation cover that generates the clouds that produce the rains'.[64]

I needed to bring to the attention of the court all this important information. I needed to stress what was really at stake when destroying a cloud forest, and why destroying a cloud forest would ultimately have a crucial impact on water sources for the world as a whole.

Levy Ortiz pointed out that 'the montane forests of Ecuador, as in the case of the Los Cedros Protected Forest, are particularly important for the water cycle in an area much larger than the one they cover, due to the capture of water through their high diversity of plants epiphytes, such as ferns, bromeliads, and orchids that live on trees. These epiphytes absorb water from fog, helping these forests to capture up to 75 per cent additional water through fog, allowing cloud forests to maintain a steady flow downstream during dry periods.'[65]

I learned that, by 1970, cloud forests covered about 50 million hectares of the Earth's surface.[66] Now, montane and upland forests were disappearing at a faster rate than any other tropical forest. Greenhouse gas emissions is one of the factors causing this.

I was rapidly learning, through the Los Cedros case, how everything is connected, interdependent. But Western thought has known it all along. When Alexander von Humboldt, the eighteenth-century naturalist and explorer, went to the Americas in 1799, he noted in his diary: 'Everything [in Nature] is interconnectedness' (*Alles ist Wechselwirkung*).[67] He had come to the realisation that Nature (organic and inorganic matter in Nature) was a web of intricately entwined elements, in which humankind was just one small part, and in which even the tiniest of creatures had a role to play.[68] Humboldt saw the inner, organic life of plants, animals and humans in the same plane, and found an interconnection between the smallest bit of moss and the tallest mountain.[69] He called this unity of all '*Naturgëmalde*', which in German means 'unification'.[70] In *Cosmos*, his magnum opus, which took him nearly twenty-five years to write, he delved into the interconnectedness of nature and being.[71]

In short, Humboldt revealed Nature to the West as a single interconnected organism. When did the West forget this?

Going back to my work on Los Cedros, centuries away from Humboldt, I continued my 'exploration'. 'Climate change affects cloud immersion in two major ways', explains Eileen Helmer, a research ecologist. 'First, it causes ambient air temperatures to become warmer, in turn forcing clouds to form at higher, cooler elevations. Second, it reduces humidity, which causes thinner and less frequent clouds – meaning that when clouds do condense at higher elevations, there will be less of them.'

'When this happens', Helmer says, 'many species are going to be lost.'[72]

The law protects Los Cedros from it all

In his book *Awe: The Transformative Power of Everyday Wonder*, Dacher Keltner, a professor of psychology, makes the point that happiness comes down to finding awe.[73] Awe, defined as the capacity of 'being amazed at things outside yourself',[74] allows us to discover 'the deep partners of life'.[75] 'We have a basic need for awe wired into our brains and bodies', posits Keltner.[76] We have a 'biological need for wild awe'.[77]

At work on the Los Cedros case, I was in constant awe. As I worked through a heatwave, under lockdown, I went through vast amounts of scientific evidence and would become so subsumed with my task that I lost track of the hours as I was drafting. I retain a recollection of working uninterruptedly with laser-like focus for hours, listening to Bowie's *Starman* at full volume as I did so. I was trying to do all I could to save an irreplaceable world.

My *amicus* raised Ecuador's obligations under the 1992 Convention on Biological Diversity, the Aichi Targets, the 1992 United Nations Framework Convention on Climate Change, the Paris Agreement, the 1972 UNESCO Convention concerning the protection of the world's cultural and natural heritage, as well as regional agreements specific to the Americas, all of which are directly justiciable in Ecuador's legal system. But the law does not exist in the abstract. A major effort in my endeavour was to present before the Court the actual facts – facts that neither the government nor the private sector had presented to the court.

Those many hours of delving into the details proved to be worthwhile. In a conference, many years later, the presiding judge, Agustín Grijalva, stated 'some American universities have observed that our judgment reads more as a treatise on biology than as a "legal judgment".' He shrugged his shoulders and continued:

'Understanding the scientific aspects in this case was crucial for reaching a verdict. This was the voice of nature.'[78] A colleague who saw the submissions I had presented, once the court made them available, told me: 'Your submissions clearly influenced the manner the Court approached the subject, it followed your structure giving importance to the facts demonstrated by the science.' If I managed to help the Court in any way, I was proud that this had been my work.

The mining project being the forest's most imminent threat, I concentrated on that particular point. The mining project had argued that Los Cedros was not intangible but only protected under secondary legislation (i.e. not the constitution) in Ecuador. If so, I argue, this was in clear violation of constitutional and international law. What matters is the biological and genetic value of Los Cedros, evidenced by the science. This is protected by the law that trumps any lower-ranking legislation.

Article 73 of the Ecuadorian Constitution provides: 'The State will apply precautionary and restrictive measures for activities that may lead to the extinction of species, the destruction of ecosystems or the permanent alteration of natural cycles.' Article 14 of the constitution recognises, on the other hand, biodiversity as a strategic sector in Ecuador: 'The preservation of the environment, the conservation of ecosystems, biodiversity and the integrity of the country's genetic heritage, the prevention of environmental damage and the recovery of degraded natural spaces are declared of public interest.'

This was picked up by the Constitutional Court. Even if you were to be in doubt about the invaluable worth of Los Cedros, the Court was bound to take the precautionary approach.

The diversity of life

I filed my nearly fifty-page *amicus* on 11 August 2020,[79] and, some months later, on 19 October, I attended the hearing set up by the Court. The list of oral interventions, including *amicus curiae* (non-party) interventions, was long. Each lawyer was allocated some time.

This was a courtroom of the future. A courtroom where the key alleged injured party was Nature itself.

The hearing started in the afternoon (London time) and I was hoping that I would stay focused for the time when it would be my turn to address the court. Sitting there for hours, in my gown and wig, I was taking profuse notes in order to make pertinent submissions when my time arrived.

The oral submissions did not comprise the reading of any previously submitted statements. It was entirely oral, not following any script. First the Parties, then 'the friends of the court' or *amicus*. Some of the scientific interventions used slides, photographic material and other aids, focusing on different valuable species in Los Cedros: on fungi, on frogs, on the Andean Bear, on orchids, on the myriad species existing in Los Cedros. 'Every species is a magic well', as Edward O. Wilson would say.[80]

I realised that my role had been to bring together each of these specific scientific interests into one single document, as if writing the symphony all these instruments created; summing up the full scientific value of Los Cedros. I had relied on many scientific papers published on the topic; data that had now become evidence in this case. I had then subjected this to a legal analysis that left no stone unturned.

Then, Cornerstone took the floor. We were already some hours into the hearing, but my alertness suddenly sharpened. Cornerstone warned the Court that its mining project was protected by a trade agreement between Ecuador and Canada, and an outcome negatively affecting that investment would expose Ecuador to an investment arbitration, to be filed by Cornerstone. You could hear a penny drop, so silent had the courtroom become.

Luckily, my turn came after Cornerstone. I took the floor and, among others, dealt with this point. 'Mr President, Honourable members of the Court, even if one were to look at this case from the prism of investment law, it is the investors' obligation under said law to do due diligence prior to investing, and surely Cornerstone is familiar with the Ecuadorian Constitution, which protects species from extinction and protects biodiversity preservation – International Investment Law does not support the breaching of the Constitution of the host State,' I submitted. My legal instinct told me that this point was an important one to quickly put to bed, so I did.

During the hearing, the judges made a number of interventions themselves, posing questions to the parties intervening. The hearing ended very late London time. It had lasted for about eight hours in all. Had that just happened? Had I been advocating for a non-human entity, monkeys, bears, fungi? I thought next morning, on waking up.

While we waited for the judgment to come in the following next months, I realised that the underlying legal issues had caused me to reflect deeply, well beyond the case. While Ecuador has a non-anthropocentric constitution, for the rest of the world things are different.

The 'Human Rights' paradigm

The Incas, my ancestors, used the word 'Pachamama' to denote the Earth in Quechua, the language of the Andean civilisation. To them, the Pachamama is a creative power that sustains *all life* on this Earth. In the world of my ancestors there was a respect for the

Earth, which I have seen mirrored in indigenous cultures for whom I have acted in other parts of the world: the Wayúus, the Wiwas, the Kogis, the Arhuacos, the Maswals, the Cho'rti's, the First Nations in Australia and the Pacific. It is the respect for all living entities. Before planting seeds, one 'asks the Earth for permission'. The landscape contains the soul and the roots of the people.

With the advent of human rights reflected in the 1948 Declaration of Human Rights, the right to life as a legal notion became an anthropocentric notion. It somehow marked the loss of the indigenous visions of the world, in which all the living had equal rights, or – more than this – mountains, rivers, the earth were given divine status.

Did we get it wrong? Has a system of law that only respects the right to life of human beings (legally speaking) missed the true protection of *life*? In the midst of climate change and Covid-19, and faced with the collapse of the natural world around us, are we finally waking up to the narrow-mindedness of such an approach?

The development of binding treaties, such as the International Covenant on Civil and Political Rights and the International Covenant on Economic, Social and Cultural Rights, marked an important recognition of universal human rights but, at the same time, an unfortunate separation of human from environmental rights.

Environmental rights developed later, properly speaking from 1972 onwards, and by means of non-binding declarations.

This unfortunate separation of what constitutes human rights and what is the environment is now reaching a critical reversal, which started with international courts like the Inter-American Court on Human Rights acknowledging, in 2017, the intrinsic relationship of human life and a healthy environment.

Just a few months after I submitted my brief in the Los Cedros case, the European Court of Human Rights held a high-level conference on 'Human Rights for the Planet'. I was privileged to participate. It was a reflection on the Strasbourg body of law and its suitability to protect human life in its inextricable link to the environment.

It felt as though we were at a turning point – something similar to what our elders may have experienced during the Enlightenment.

A change of paradigm

A non-anthropocentric perspective is not unknown to international law.

In between the 1972 Stockholm Declaration, which initiated a process of normative development[81] in international environmental law, and the 1992 Rio Declaration (both anthropocentric), the states comprising the UN General Assembly adopted, in 1982, an Earth-centred instrument that contains, in my view, the seminal agenda for an altogether different direction in the development of humanity. It declares that 'Nature shall be respected and its essential processes shall not be impaired'.[82]

Why has this been forgotten?

This was the 'World Charter for Nature', which set forth 'principles of conservation by which all human conduct affecting nature is to be guided and judged'.[83] The Charter was adopted by a vote of 111 states in favour, including developed states like the United Kingdom. It has been held to be 'an avowedly ecological instrument, which emphasises the protection of nature as an end in itself'.[84] When I read this forgotten declaration, it appeared clear to me that, at some point between 1982 and 1992, we had lost our way and we needed to retrace it.

In an illuminating way, the preamble of this declaration noted that: 'Mankind is part of nature and life depends on the uninterrupted functioning of natural systems', that 'every form of life is unique, warranting respect regardless of its worth to man' and that 'lasting benefits from nature depend upon the maintenance of essential ecological processes and life support systems, and upon the diversity of life forms, which are jeopardised through excessive exploitation and habitat destruction by man'. Isn't this in accordance with those immutable laws that indigenous peoples also respect? Aren't these the principles that should be guiding our human laws?

We may be standing at the very crucial moment where notions such as our idea of *progress* and *development* need to be reconceived, reconsidered and a new paradigm embraced. Perhaps the notions at the heart of the Constitutional Court of Ecuador in the Los Cedros case may serve to advance our understanding of such a paradigm and may be followed as an example by other systems of law.

Should trees have standing?

Such a paradigm shift would stop us seeing Nature as 'natural resources', or something to 'exploit'. Naturalists have observed that humanity has treated Nature as 'something out there – nameless and limitless, a force to beat against, cajole and exploit.'[85]

It is often noted that a 1972 book, called *Should Trees Have Standing*,[86] which explored the possibility of having nature as a plaintiff in the United States, would have initiated the Rights of Nature vision. In reality, however, the notion that Nature is an entity deserving respect by humans and endowed with life is not an idea that has come from post-industrialised societies. It is a vision that has existed for millennia – with indigenous peoples.

While *Should Trees Have Standing* is an important reflection from the perspective of a Western common-law system and the possibilities within its procedural law, I do not consider that this book provides any model for a Rights of Nature paradigm shift. It is written from a perspective that continues to look upon Nature as 'natural resources',[87] capable of being monetised, thereby not seeing Nature for its own intrinsic value.[88] It does not challenge the underlying economic model, which requires such 'exploitation' and its correlative notion of 'development' and 'progress'.[89] It is also a book that makes indigenous peoples' visions on Nature invisible. If you are interested in a true understanding of a system of law that is anchored in universal laws, turn to indigenous peoples' knowledge. Turn to indigenous peoples' rights law. Its rich jurisprudence shows that such visions are part of international law, albeit often violated.

The premise upon which the new paradigm ought to be erected should listen to the science and to the indigenous peoples' cosmovisions – just as the example of the Torres Strait Islanders' case shows to be possible. This is a paradigm that would elevate the very concept of life.

The ruling of the Constitutional Court of Ecuador

On 1 December 2021, we were notified of the Constitutional Court of Ecuador's judgment in the Los Cedros case.[90] I did not know what to expect. I had done my upmost, but it is never entirely clear how a particular court may decide a case.

I opened the notification with some apprehension, with the same expectation we open defining personal news coming in the mail. As I read the judgment, I felt all the blood in my body going towards my head, as I tried to rapidly comprehend the details. Then I realised: victory! Los Cedros had been vindicated. This was a first decision of its kind on the Rights of Nature, and it was a landmark judgment.

In the judgment, delivered in Spanish, the Constitutional Court of Ecuador had prioritised Nature above economic interests. It had given more importance to life itself.

As Judge Agustín Grijalva reflected at a conference I attended in January 2024: How much are the lungs of the Earth worth? How much the species that are going extinct?

The Court held that the Rights of Nature, enshrined in Ecuador's Constitution, had 'full normative force'.[91] It emphasised that the rights of Nature were not merely ideals or rhetorical statements but entailed 'binding legal obligations'.

Hinting at the lack of consideration of these rights by the authorities that had granted the concessions, the Court noted that these rights were often not 'timely and adequately considered by some judges, other public and private authorities'.[92]

The presiding judge, Agustín Grijalva, would observe later, during an interview many months after delivering the judgment, that what he found somewhat surprising was that the agencies of the State had not provided any assessment of the scientific value of the forest to the Court. They had not carried out any assessment of this prior to granting the concession.

In its judgment, the Court developed jurisprudential principles (both substantive and procedural) in relation to what is understood as 'Rights of Nature' under Article 71 of the Constitution of Ecuador. The 'intrinsic value of Nature' was elaborated to mean a 'systemic perspective that protects natural processes for their own value'.[93] 'In this way', the Court observed, 'a river, a forest or other ecosystems are seen as life systems whose existence and biological processes merit the highest possible legal protection that a constitution can grant: the recognition of rights inherent to a subject.'[94]

This brought the natural world into a different ontological order; as deserving 'the highest possible legal protection' in a legal system.

The *Los Cedros* judgment conceded that it is difficult to understand the intrinsic value of nature through the recognition of rights, if one is to look at the world from a rigidly anthropocentric perspective, 'which conceives the human being as the most valuable species, while reducing the other species and the nature itself, to a set of objects or resources to satisfy human needs, especially those of an economic nature'.[95]

'It is false that the current law in the world is anthropocentric,' stated an indigenous man who spoke very assertively on this point at a conference I once attended. 'The world is capital centric [. . .] multinationals have more rights than any human being.'

This was not a casual remark. It was the remark of a member of the *pueblos originarios*, indigenous peoples from Latin America who had experienced this at first hand. Gabriel García Marquez depicted this reality too. A passage from *One Hundred Years of Solitude* comes back to me. A banana company had modified the natural environment, 'moving the river from where it had always been', just as Glencore, Anglo American and BHP had done in the territory of the Wayúu:

> There was not much time to think about it, however, because the suspicious inhabitants of Macondo barely began to wonder what the devil was going on when the town had already become transformed into an encampment of wooden houses with zinc roofs inhabited by foreigners who arrived on the train from halfway around the world, riding not only on the seats and platforms but even on the roof of the coaches. The gringos, who later on brought their languid wives in muslin dresses and large veiled hats, built a separate town across the railroad tracks with streets lined with palm trees, [. . .] they changed the pattern of the rams, accelerated the cycle of harvest, and

> moved the river from where it had always been and put it with its white stones and icy currents on the other side of the town, behind the cemetery.[96]

In the bizarre manner in which fiction evokes non-fiction, García Marquez portrayed the world denounced by the indigenous man I referred to above as 'the utilisation of nature for capitalist greed', going as far as 'relocati[ng] . . . the river to irrigate the banana plantations'.[97]

As in the world of my clients, Nature is a character in García Marquez' work, a natural world that the '*gringos*' (corporations) and 'progress' appear to change, destroy, cause to disappear. In *The General in His Labyrinth*, a character (Bolivar) faced with the pollution of the Magdalena river, brought by 'civilisation', declares: 'The fish will have to learn to walk on land because the waters will end.'[98] This literary prophecy sounded disturbingly real rather than magic realism to me. Rivers around the world that I encountered in my work are dying. Could reverting to the respect for Nature overturn this destruction?

Even the argument of 'economic development' and 'progress' seemed not to be quite straightforward. In the Ecuadorian example, in order to show that the company created employment for the locals, a mining company in an area not far from Los Cedros had been paying people to do nothing, a local noted. This was simply to gain some support, even if it was not creating any real development for anyone. People saw this as 'easy money' and were corrupted into accepting the mining project.

But at the *Los Cedros* judgment, in contrast, the Constitutional Court of Ecuador asserted a new paradigm that no longer saw Nature as a natural 'resource'. The Court stressed that: 'It is about a change of legal paradigm because historically the Law has been functional to the instrumentalization, appropriation and exploitation of nature as a mere natural resource. The Rights of Nature propose that in order to harmonise our relationship with it, it is the human being who adequately adapts to natural processes and systems.'[99]

I felt that the cosmogony of the original inhabitants of the Americas was being vindicated in this dictum. The deep Andean vision reflected in Gamaliel Churata's work, which went back to the origin of it all, the genes, the basic physical and functional unit of heredity, the DNA of the natural world, was being vindicated in the interpretation of the Court. Under this law, activities that lead to the extinction of species constituted a violation of the Right of Nature to have its existence fully respected.[100]

This may appear common sense, but ask your own legal system whether it protects species from extinction. The Ecuadorian Court held this to be:

> a violation of such magnitude that it would be equivalent to what genocide means and implies, in the field of human rights. Once a species is extinct, the laborious process that has sometimes taken millions of years for nature to produce results in an irreparable loss of diversity and knowledge. Precisely because of the serious and irreversible damage such as the extinction of species, Article 73 of the Constitution applies the precautionary principle to these cases.'[101]

In addition to being a violation of the Rights of Nature, such violations, in the view of the Court, could have unsuspected negative effects on human beings, which would also violate other rights, such as the right to water and a healthy environment.[102] Of course indigenous people will say: 'We are Nature.' What hurts the environment, hurts us. The separation of Nature from mankind exists only in the minds of Western civilisation. It should not. Already in the seventeenth century, more than a century before Humboldt, writing in the Netherlands, the philosopher Baruch Spinoza posited in his *Ethics* (1677) the human being as part of Nature. He wrote: 'the laws and rules of nature, according to which all things are made and change from one form into another, are everywhere and always the same, and therefore, one and the same manner must there be of understanding the nature of all things, that is, by means of the universal laws and rules of nature.'[103] During his time, he was persecuted. Today he is considered one of the great thinkers of the European philosophical tradition.

Are his ideas so forgotten today?

Instead of following Spinoza's approach, there is a blind Benthamian (as in Jeremy Bentham's) utilitarianism to much of our thinking and I do believe this intellectual discrepancy has a bearing on the way society sees the world. The West has adopted a singular way of looking at things, but it's contingent on fundamental, first principle assumptions coming from such a perspective. There's no reason we have to think in the way we do – and to interpret the law in a way that has rethought those first principles is at the heart of this book.

Away from such non-existent divisions of humans and Nature, in Ecuador, in Los Cedros, in the twenty-first century, we were making legal history. As a result of its findings, the Constitutional Court of Ecuador declared the mining project and concession in Los Cedros to be in violation of the Rights of Nature, as well as in violation of the right to water and the right to a healthy environment of communities around the forest. Consequentially, the Court ordered the mining enterprises responsible for this to 'abstain from any activity in Los Cedros'[104] and to fund the restoration of what has been damaged so far, as a consequence of said activities. This legal consequence is unprecedented in legal history.

I believe that a decisive battle regarding the protection of endemic species and natural habitats, such as primary forests, was won in Ecuador with this judgment, thereby setting a world precedent.

I understood that it was important for the world to know and to follow such important developments in Ecuador. So, before the hearing of the Los Cedros case, I had already contacted a media entity that helped me spread the news across a number of newspapers in London and beyond. I felt it was important the world knew what was at stake and that Ecuador knew that the world was watching. *Le Parisien* published a piece entitled 'Espèces menaces: il faut sauver l'ours Paddington' ('Threatened species: We Must Save the Paddington Bear'). The Andean bear, one of the species threatened in Los Cedros, was known in Europe as the 'Paddington bear', because of the well-loved English children's books. Applying my skills as a litigator and 'storyteller', I had thought this was the angle that might capture the imagination and be likely to pick up the interest of some newspapers. I was not wrong.

When the judgment came out, we made sure that the news was propagated across the world in English, Spanish, Portuguese, Polish, French, Thai – from Paris to Melbourne. One morning I got up to see a wonderful piece by Jane Dalton in *The Independent*, which featured some of the species that had been saved from extinction. I rejoiced. I was interviewed by *The Ecologist* and I told them that, to me, Los Cedros was 'the case of the century'.

Can there be magic in the law?

When Judge Grijalva reflected on why the case of Nature won, he acknowledged that the advocacy work around the value of the forest had moved consciousness. The Court had received twenty-seven *amicus curiae* in the case.

The *Los Cedros* decision gives hope. It is one of the most significant decisions to arise on the Rights of Nature worldwide, to date, at a time when biodiversity loss and climate change are of growing concern. This dictum (of nearly a hundred pages) is an important contribution to the growing jurisprudence on the protection of the natural world. The jurisprudential reasoning on the right to water and the right to a healthy environment adds up to developing jurisprudence in those areas of law. Doubtless, this dictum will also inspire other jurisdictions that are considering adding similar precepts in their own constitutions.

TEN

Into the Deep Blue

In a scene from *Star Trek IV: The Voyage Home*, an unidentifiable sound puzzles Admiral Kirk and Dr Spock. Kirk asks, 'Spock? What do you make of that?' And Spock answers, 'Most unusual. An unknown form of energy of great power and intelligence.' Spock eventually identifies that these are the songs of humpback whales. Admiral Kirk then states: 'Whales have been on Earth far earlier than man,' and Dr Spock observes: 'Ten million years earlier. And humpbacks were heavily hunted by man.' He then adds: 'They've been extinct since the twenty-first century.'

Kirk and Spock were in a fictional future, where the songs of whales were voices of extinct creatures. The sad thing is that this resonates no longer as science-fiction but as a pressing danger, due to the climate breakdown, which reason, including legal reason, should be able to avert.

A meeting with a sovereign and a trip to Katowice

When the foreign state official entered the room, I was ready to give my opinion. It was the end of 2018 and he represented a sovereign state. He wanted to know what my opinion was in relation to a potential case before the International Tribunal for the Law of the Sea, addressing climate change. This is a Tribunal that deals with inter-state matters and the law of the sea.

The state official was interested in knowing whether a climate change claim under the Convention on the Law of the Sea, the multilateral treaty whose implementation the Tribunal oversaw, was viable. He also wanted to know whether it was better to seek the jurisdiction of the Hamburg court or the International Court of Justice (or World Court) in The Hague.

Unlike the World Court in The Hague, which is often in the news, the International Tribunal for the Law of the Sea, consisting of twenty-one judges, is more obscure and, until recently, little-known by the media. But this court had the potential to play a key role in addressing climate change because of climate change's impact on the oceans. About 70 per cent of the Earth's surface is comprised of ocean.[1] The ocean is critical for all life on Earth.

'I don't see how the Convention on the Law of the Sea can address climate change,' another barrister had observed in a previous informal meeting. I had become impatient. He was clearly no expert on the law of the sea, but sometimes the overconfidence of male lawyers who lack expertise is astounding. 'It is a living instrument!' I retorted. 'Of course the term pollution in Article 1 of this treaty can be interpreted to include greenhouse gas emissions,' I added. This was obvious to me.

I had previously worked at the International Court of Justice and was familiar with the work it undertook. Now, in the meeting with the representative of the foreign state, the state official turned to me and asked me for my opinion. 'What do you think?' he asked as he looked at me. He was interested in knowing whether a climate change case could have a better chance before the Tribunal for the Law on the Sea or the World Court.

The barrister with no expertise regarding the law of the sea interrupted. He cut me short in order to give his opinion, even though he had not been asked and had never worked, appeared or had any previous experience with the procedure of the International Court of Justice.

I realised that I needed to bring in another expert, a specialist in Public International Law, like me, a barrister with the highest expertise possible in relation to the law of the sea, to provide advice moving forward. And so it was that I called up Professor Alan Boyle, also a barrister, and an eminent expert on environmental law and law of the sea. Professor Boyle agreed with me, that the Convention on the Law of Sea was fit to tackle climate change. He graciously agreed to provide advice with me in an upcoming meeting with a group of states in Katowice, Poland, where COP24 was going to take place in December 2018.

Katowice, the capital of Upper Silesia in Poland, was cold. In fact, it was snowing. But there we were, in this city whose predominant industry had been the production of coal, with a purpose. Professor Boyle had arrived after driving all the way from London. We had something to eat, and then we addressed the states interested in this topic, in a closed session to which we had been invited.

I was hoping that, by December 2018, our presentation would precipitate some action. But things moved at their own pace. In retrospect, I should not have been surprised. Not even the non-governmental organisations (NGOs) were persuaded that one could seek climate action through international courts back then. We were wasting valuable time. And I was impotent to do anything about it.

I did not stay in Katowice but rather in Kraków, a beautiful city a short distance away, where I sought some time off, watching silent movies with live music.

Then, towards the end of 2019, I was formally instructed by a sovereign state (the country of the same state official who had come to visit me in chambers) to provide, alone, formal written advice on the viability of an Advisory Opinion on climate change by the Tribunal for the Law of the Sea. In my legal opinion this was viable. I sketched the procedural route such a request could take and the relevant questions one could ask the Tribunal to address.

It took three years more, however, for some states to take the initiative to seek a pronouncement by the Tribunal for the Law of the Sea on the obligations of states in relation to climate change and the oceans.

At last, in a request dated 12 December 2022, the Commission of Small Island States on Climate Change and International Law (COSIS) (which Professor Boyle happened to be advising) asked the International Tribunal for the Law of the Sea for an Advisory Opinion on two questions. First, 'what are the specific obligations of State Parties to the UN Convention on the Law of the Sea, to prevent, reduce and control pollution of the marine environment in relation to the deleterious effects that result or are likely to result from climate change?' The second question was: 'what are the specific obligations of States to protect and preserve the marine environment in relation to climate change impacts?'

The questions posed in the Request by COSIS contained the underlying assumption that anthropogenic greenhouse gas emissions are a form of pollution of the marine environment, as they cause deleterious effects, just as I had argued in 2019. The procedural route COSIS took to lodge the jurisdiction of the Tribunal was the route that, in essence, I had suggested in my advice in 2019. The Tribunal received this Request and set out a deadline of 16 June 2023 for submissions of state parties.

The answers the Tribunal was set to provide are key for the future of the ocean and the planet. At the centre of the questions was the issue of whether this old treaty can be interpreted in the light of new developments, mainly polluting activities that have emerged since and that are currently harming the oceans.

A barrister for the oceans

An 'Advisory Opinion' is an international court's interpretation of the law delivered following formal court proceedings before the International Tribunal. The Tribunal for the Law of the Sea has two types of jurisdictions: a contentious jurisdiction, and an advisory one. Single states, even if they are parties to the Convention on the Law of the Sea, cannot trigger the Advisory Jurisdiction of the Tribunal alone. The Rules of the Tribunal[2] state that the Tribunal may give an Advisory Opinion on a legal question if an international agreement related to the purposes of the Convention specifically provides for the submission to the Tribunal of a request for such an opinion. An 'international agreement', in my view, includes bilateral agreements. In any event, an international agreement referred the questions on climate change to the Tribunal on the Law of the Sea for an Advisory Opinion. The proceedings follow a written stage of submissions and an oral stage. While Advisory Opinions are not legally binding, they are authoritative pronouncements by a court or tribunal, which then guide state action.

The proceedings I envisaged in 2018 had finally reached the Tribunal for the Law of the Sea! This was an unprecedented and defining case.

In May 2023, I was instructed by the World Wide Fund for Nature (WWF) International, the world's leading independent conservation organisation, to act as their counsel and prepare court submissions to file in these proceedings, during the written submissions phase of the case. The WWF's submissions in the past had been influential upon the outcome of other previous Advisory Opinions. This was because of the scientific expertise it held.

Drafting submissions for the Tribunal kept me busy during the summer of 2023. I was subsumed in the universe of krill, whales, polar bears, water columns, the Antarctic and the deep sea. My submissions were covering a lot of technical ground, but in essence they were concerned with wide-ranging topics (i.e. spanning Arctic ice and deep-sea mining), the impacts of climate change and what they meant for the law of the sea.

The WWF have a wealth of scientific knowledge on the impacts of climate change on marine ecosystems, and I was working with their scientists to bring these submissions into shape. It was a phenomenal effort.

I was aware, of course, of the general impact of climate change on the ocean, but until I delved into this case I had not realised just how precarious the situation was.

The Intergovernmental Panel on Climate Change (IPCC), the UN body tasked with assessing the science on climate change, has noted that 'ocean and coastal ecosystems support life on Earth and many aspects of human well-being. Covering two-thirds of the planet, the ocean hosts vast biodiversity and modulates the global climate system by regulating cycles of heat, water and elements, including carbon.'[3] Due to its size and reflective capacity, the ocean 'has absorbed more than 93 per cent of the heat generated by anthropogenic global warming since 1971'.[4] In effect, the ocean is being used as the planet's greatest carbon sink, absorbing around 90 per cent of the excess heat and energy released from rising greenhouse gases[5] and around 20–30 per cent of the carbon dioxide (CO_2) produced by human activities since the 1980s. This has resulted in changes to ocean chemistry that are unprecedented in 65 million years.[6]

The IPCC identifies that climate change has already caused 'substantial damages, and increasingly irreversible losses in terrestrial, freshwater, cryospheric and coastal and open ocean ecosystems'.[7] The 'cryospheric' is the frozen water part of the Earth's system. It includes frozen parts of the ocean, such as the waters surrounding Antarctica and the Arctic.[8]

The IPCC has identified widespread impacts and related losses and damages to human systems and altered ocean ecosystems worldwide. Warmer waters have led to shifts in marine species' ranges as they attempt to adapt to new conditions including prey availability – some fish populations are projected to be pushed predominantly towards the poles, leading to some local extinctions in the tropics and invasive species at higher latitudes.[9] As the IPCC observes, 'hundreds of local losses of species have been driven by increases in the magnitude of heat extremes' and 'mass mortality events on land and in the ocean'.[10]

What particularly drew my attention was the realisation that the impacts on some ecosystems are approaching irreversibility,[11] 'such as the impacts of hydrological changes resulting from the retreat of glaciers, or the changes in some mountain or Arctic ecosystems driven by permafrost thaw'.[12] As someone told me: 'You can plant trees, but you cannot plant ice.'

Sitting in my kitchen, I read the IPCC report of 2023. I think best early in the morning, so I often start at 5 a.m., sometimes 4:30 a.m. It is summer and I have all the light of the world. I am absorbed by what the science states, trying to figure out what this means for the law; its translation into state obligations.

Through the French doors of my kitchen, with all my references spread out on the table, I could see an unexpected visitor landing by the door. It was a great spotted woodpecker with a vivid red patch at the back of the head, coming to say hello. His visit may have inspired me to start the submissions on behalf of WWF with a paragraph reflecting on the Rights of Nature: 'This brief is written from a perspective whereby a healthy marine environment [. . .][13] is important in and of itself and not merely for the service it renders to humanity. The UN Convention on the Law of the Sea protects the marine biological diversity of the oceans and therefore the right of the diversity of marine species to exist and thrive.'

The law supports such a notion. That the 'marine environment' is not a solid object, like a sort of car park, devoid of life, but it is something living, moving, breathing. It is constituted by organisms. We have all come in contact with the living sea. Yet we know so little about it. I am particularly intrigued by the biodiversity in the deep sea, which we don't know much about, but I have been trying to understand better. Each creature I encountered in my readings is a subject of fascination. My Twitter/X followers, most of them lawyers, must be surprised at my tweets recording the strange deep-sea creatures I encounter, but I am interested in the millions of invertebrate organisms. 'Often overlooked', they form 'the foundations of Earth's ecosystems.'[14]

Melting ice: The poles

The warming of the waters due to emissions of greenhouse gases into the environment is upsetting the order of natural things. The food that seals, sea lions, whales and dolphins depend on has been shown to change locations due to warming water. The loss of sea ice and associated abundance of prey species is also impacting terrestrial species, such as polar bears. The IPCC has observed in that regard: 'The polar regions are losing ice, and their oceans are changing rapidly. The consequences of this polar transition extend to the whole planet and are affecting people in multiple ways.'[15]

'[I]t is *extremely likely* that the rapid ice loss from the Greenland and Antarctic ice sheets during the early twenty-first century has increased into the near present day, adding to the ice sheet contribution to global sea level rise', reads the IPCC report.[16] It is further noted that at sustained warming levels (between 2°C and 3°C), the Arctic Ocean will soon be practically sea ice free throughout September in most years. There is also limited evidence that the Greenland and West Antarctic ice sheets will be lost almost completely and irreversibly over multiple millennia; both the probability of their complete loss and the rate of mass loss will increase with higher temperatures.[17]

Critically, the Arctic is warming nearly four times as fast as the global average, and loss of land ice is causing sea level rise, threatening coastal communities and existentially threatening marine species and sinking island states. The ice that polar bears depend on is melting away. Loss of ice is also threatening other species, such as seals, who depend on this habitat to raise their young.[18]

At the end of the food chain in the sea is the krill, mainly found in the Antarctic. A tiny shrimp-like crustacean, it is the primary food source for blue whales and some seabirds.[19] As reported, '[w]hen abundant, animals migrate thousands of miles to feed on krill. But when absent the entire marine ecosystem suffers.'[20] Global warming is now depleting the sea ice that krill depend on. Scientific research suggests that warming waters will impact krill growth, 'possibly leading to a 40 per cent decline in the mass of individual krill by the end of the century',[21] with further research arguing that 'ocean acidification, another impact of climate change, will reduce krill development and hatching rate and lead to an eventual collapse in 2300.'[22] The blue whale would go extinct well before this.

I caught my breath. The dialogue of Kirk and Spock in an imaginary future that I opened this chapter with was starting to sound dangerously plausible. It's likely to occur within my life span if things continue as they are.

'We are driving a whole environment to extinction', remarked Robbie Mallett, a sea expert at University College London, Earth

Science, speaking to the *Guardian*.[23] 'Even if planet-heating emissions are radically cut, the world's vast ice sheets at the poles will continue to melt away for hundreds of years, causing up to three metres of sea level rise that will imperil coastal cities', observed the *Guardian* based on a scientific report released at COP27 in 2022.[24]

Could the Tribunal's work make a difference to mitigate this?

I believe it can and is on the way to doing so. If the Tribunal provides a reply to the two questions with which it was presented, accepting that greenhouse gas emissions constitute pollution of the marine environment, and clarifies the obligations of states in preventing this, requiring some rapid steps in that direction, a major change will have to take place around the globe. The Convention on the Law of the Sea is highly ratified by states. The jurisdiction of this Tribunal has a potentially wide-ranging impact.

When, in 2005, the Inuits brought their case (dismissed with short-sightedness by the Inter-American Commission), Sheila Watt-Cloutier, the petitioner, was asked by a journalist, 'what do you most want people to understand'? She replied: 'Think about the interconnectedness of the arctic ice. What happens in the arctic doesn't stay in the arctic. It's impacting the rest of the planet. The arctic is the air conditioner for the world, and it's breaking down.'[25] Unfortunately it has taken eighteen years for the world to start acting.

As often happens when drafting, I reflected overnight on my court written submissions. I got up early and did some work in the garden while reflecting on the draft, my thoughts crystalising. The topiary in my walled garden had been invaded by invasive species of caterpillars that have to be removed by hand (if one wants to avoid using poison), so there I was, plucking caterpillars from box bushes and reflecting on the draft.

Something was nagging at me. Everybody seemed to agree that the standard of conduct in protecting the marine environment was one of due diligence. Yet what the best available science today demonstrates is that some specific, rapid steps are required. The Convention on the Law of the Sea requires states to take measures necessary to prevent, reduce and control marine pollution 'from any source'.[26] The burning of fossil fuels is the largest cause of this pollution. The IPCC's most recent reports have made it clear that greenhouse gas emissions from existing fossil fuel infrastructure will already push the world beyond 1.5°C of global warming, which, as we saw above, is rather dangerous for the world. Therefore, as per the obligations under the Convention on the Law of the Sea, of 'reducing' pollution, states are expected not to grant licences to, but rather to prevent, new oil and gas exploration. This was an obligation to 'refrain from' a specific conduct. This, I submitted, was an obligation of result (an obligation that requires the State to ensure the obtaining of a specific result). In short, states have the obligation to refrain from granting new fossil fuel exploration under the Convention on the Law of the Sea.

The deep sea

The sea floor is one of the mostly untouched ecosystems in the world, but it has been at the centre of a sort of 'gold rush' in recent years. The seabed beyond national jurisdiction ('the Area'),[27] which covers approximately half of the planet,[28] contains deposits of valuable minerals, such as cobalt, manganese, copper, zinc, nickel and rare earth elements. These are particularly sought after by green technologies (including electric vehicles, wind turbines and solar cells) and tech-oriented industries, such as the smartphones industry.[29]

The International Seabed Authority (ISA), an intergovernmental body established under the Convention on the Law of the Sea, based in Kingston, Jamaica, has awarded thirty-one exploration contracts involving twenty-two contractors[30] so far. Exploration, however, may soon give way to exploitation. In June 2021, Nauru – a tiny island country in the Pacific Ocean – notified the ISA of the intention of NORI, a subsidiary of a Canadian company called DeepGreen Mineral Corp., to apply for approval to begin mining in two years in a deep sea area between Hawaii and Mexico.[31]

The current tension between commercial interests for deep-sea exploitation (driven by the expanding world demand for raw materials) and environmental considerations is reflected in the calls for deep-sea mining to be halted, emphasising the adoption of precautionary and ecosystem approaches. An increasing number of states, including Costa Rica, Finland, Germany, Portugal, Sweden, Ireland, Brazil and Vanuatu, are calling for a moratorium. Recently, the UK joined in.

For the law of the sea expert, Gjerde, the current challenges regarding seabed mining in the Area arise from paradigms or assumptions about life in the deep sea that have turned out to be incorrect: the assumption that life in the deep sea was dull, distant, of little interest and that seabed mining therein could occur without much environmental disturbance; that the resources for exploration/exploitation were easily accessible; that the technology was right; and that 'we just needed to develop some potato-harvesting-type machines to enable this new regime go forward'. [32]

When researching the topic of deep-sea mining sometime in 2021, I also learned about further challenges with 'hydrate mining'. Gas hydrates are 'a crystalline solid formed of water and gas'.[33] Looking and acting much like ice, they contain huge amounts of methane.[34] Methane exists in huge quantities in marine sediments in a layer several hundred metres thick directly below the sea floor and in association with permafrost in the Arctic. Methane hydrate is believed to be 'the world's largest natural gas resource' and can be found in the shallow sediments of many deep ocean areas.[35] Scientific scholars note that enormous amounts of methane hydrate have been found beneath Arctic permafrost, beneath Antarctic ice and in sedimentary deposits along continental margins worldwide.[36]

Scientists warn that hydrate mining could generate sub-marine landslides, and that 'dissociating the methane hydrates would destabilise the sea floor'.[37] In the worst case scenario, they warn that 'huge packages of sediments could slide downhill, triggering powerful tsunamis along coastal areas'.[38] Some scholars suggest that the current legal framework is not adequately prepared to address transboundary harm triggered by the exploitation of offshore methane hydrates, in particular because the technology of such extraction is still at an experimental stage, and potential risks remain uncertain – and even untraceable – for cross-jurisdictional claims.[39] Add to this, the warnings in scientific studies that an increased release of methane from the oceans could accelerate climate change.[40]

How is one to resolve the above risks included in the context of climate change with an old treaty adopted in 1982, like the Convention on the Law of the Sea? I took the view that the topic of deep-sea mining was relevant to the scope of the questions put before the Tribunal overseeing its implementation.

The general obligation enshrined in the Convention on the Law of the Sea, which reads 'States have the obligation to protect and preserve the marine environment',[41] is applicable to the high seas and to the areas beyond national jurisdiction or as it is known, 'the Area'.[42] I submitted on behalf of WWF, in the proceedings before the Tribunal for the Law of the Sea, that 'in order to comply with this, in the current context of a climate and nature emergency, a moratorium on Deep Sea Mining is required with the resulting obligation of States not to mine the Deep Sea in the absence of sufficient knowledge about its role in regulating the climate, about the impact of mining the Deep Sea on climate change, and in light of the known and unknown risks.'[43] 'No deep-sea mining could at present be consistent with the Convention on the Law of the Sea', stated WWF's court submissions.[44]

If we were to destroy the deep sea, it is unknown what effects this would have and their extent. The marine deep biosphere comprises 'more than half of all microorganisms on the planet'.[45] It's where all life on Earth came from. In the absence of light and oxygen, surrounded by an inorganic environment, live species whose habitat 'appears to mark the inner boundary of life on the planet'.[46]

Curiously, however:

> 'if our species [. . .] accidentally burned the planet's surface to a crisp, [. . .] life could go on in the planet's deep biosphere [. . .] There, the microbes and the invertebrate predators would persist, shielded and unconcerned in their dark refugia, drawing their energy and substance from rocks, resistant to heat, perhaps evolving over hundreds of millions of years, eventually to reach the surface and generate a diversity of multiple-celled organisms, thence against heavy odds a human-grade metazoan. Whereby the great cosmic cycle might give intelligence on Earth a second chance.'[47]

But it does not need to come to that. And from the States' oral submissions presented at a two-week hearing in Hamburg, early in September 2023, in the all-white headquarters of the Tribunal for the Law of the Sea, I have reason to believe that the Tribunal will enlighten us; the world will finally react.

Nature beyond the Earth

'The ocean contains the history of all humanity. The sea holds all the voices of the earth and those that come from outer space. Water receives impetus from the stars and transmits it to living creatures,' observes Buckley, in her review of Patricio Guzman's *The Pearl Button* (2015).[48]

These voices would include those of whales. Have you listened to the sound of whales, recorded by biologist Roger Payne for posterity, on NASA Voyager Golden Records?[49] There is a mesmerising image of him, underwater, next to a whale, as if they are about to 'shake hands'. The recording brought tears to my eyes when I first listened to it. It is a deep sound. Like the voice of one of the last human beings to speak Kawésqar language. The NASA recordings of whales were the sounds that triggered Kirk and Spock's dialogue.

A project in the UK in Cambridge, I also learned, has adapted Mendelssohn's *The Hebrides* overture to highlight the declining whales population. Using sound to convey the enormity of biodiversity loss, the adaptation deleted notes of the original score. 'Crucially, the roughly 30,000 notes in the original music score approximate to the number of Humpback Whales there were in the sea in 1829 when the piece was written,' notes Jacqueline Gargett, writing on the project.[50] 'By 1920, two thirds of all Humpback Whales were gone.'[51] It is moving and powerful.

As with the whales, the Kawésqar have been almost wiped out. In Patricio Guzman's lyrical documentary *The Pearl Button*, he shows how the Kawésqar people, as per other tribes in Tierra del Fuego, Chile, 'lived in communion with the cosmos and with water'.[52] 'Surrounded by water, they travelled by water and ate what the water supplied. They were water nomads who lived in clans that moved through the fjords, travelling from island to island.' They thought of the sea as a relative.

For the inhabitants of Tierra del Fuego, Nature did not stop in the fjords of the archipelago. They saw the deep connection of the sea with the stars. Selk'nam people painted the stars on their bodies. According to Guzman in his film, 'For these people, the stars were the spirits of their ancestors.' The Selk'nam 'carved stones and painted their own bodies as starry skies as a tribute to their ancestors. For after death they believed they would become stars.'[53]

I had rarely thought of Nature beyond the Earth. For some inexplicable reason, my sight was fixed on the Earth's biosphere. But the inhabitants of Tierra del Fuego, in close connection with Nature, knew it. In close contact with dark skies, they lived in deep connection with celestial bodies. Perhaps that is the reason why they did not need the notion of 'God'. When the filmmaker asked a Kawésqar woman for the word in Kawésqar for God, she answered that 'there [was] no word in Kawésqar for God'.

I gained a sudden awareness of the interconnectedness of the biosphere with the cosmos and now, when I speak of Nature, I am speaking about it all: as the Earth within the wider universe. If water depends on the movement of the stars, the laws, the gravity and how our acts change or not are also aligned with those universal laws.

I saw a dark sky properly for the first time on a crispy winter night, in Surrey. I have always lived in cities, and moving to Surrey was my first experience of living in an area free of light pollution. There, on the slopes of the South Downs, I saw the Milky Way and felt for the first time that I could almost 'touch the stars'. In child-like wonderment, I could not stop giggling with happiness.

I learned, from documentarist Patricio Guzmán, that the connection between the oceans, the rivers and outer space is close. 'It seems that water came from outer space, that life was brought to earth by the comets,' says a voice off screen in *The Pearl Button*.

Intrigued, I read on a NASA site: 'The mystery of why Earth has so much water, allowing our "blue marble" to support an astounding array of life, is clearer with new research into comets. Comets are like snowballs of rock, dust, ice, and other frozen chemicals that vaporize as they get closer to the Sun, producing the tails seen in images. A new study reveals that the water in many comets may share a common origin with Earth's oceans, reinforcing the idea that comets played a key role in bringing water to our planet billions of years ago' and that 'Water was crucial for the development of life as we know it.'[54]

A historic win for the climate

Finally, on 21 May 2024, the International Tribunal for the Law of the Sea sat in court for the delivery of the Advisory Opinion on climate change. In a courtroom in this modern Tribunal in Hamburg, we are making legal history. It is a 153-page ruling that marks, in my view, a before and an after in the way international law approaches state obligations in relation to climate change. A momentum. A break-through. The Tribunal found that states have a legal obligation to protect the oceans and marine biodiversity from climate change. As I was listening to the President of the Tribunal reading the Opinion, I realised that it was the start of a new area in the protection of the marine environment from the current impacts of climate change.

The tribunal first looked into the question of whether greenhouse gas emissions (with the correlative climate change effects) fall within the meaning of 'pollution' under Article 1(1)(4) of the Convention on the Law of the Sea and it unanimously held that it did. It found that the Paris Agreement and other climate treaties were relevant for the interpretation of the Convention on the Law of the Sea, yet found that the obligations in climate instruments did not restrict the obligations under this treaty, since they were broader in various respects.

I heard this reading with emotion. Images flooded my mind. The time I stood alone in that auditorium, where I affirmed the potential of international courts' seizing jurisdiction to address climate change effects, when nobody believed this was feasible. I recalled the time I travelled to Katowice to warn states that the Convention protected the oceans from greenhouse gas emissions. The sweat and the tiredness after a long day, in a library, drafting my advice on how to reach the Advisory Jurisdiction of the Tribunal to pose questions on climate change. The innumerable days I had spent concentrated on this. Who would have thought this day would arrive? I shook my head. I felt immensely happy.

The Tribunal emphasised obligations of states to reduce greenhouse gas emissions in a number of areas, namely emissions from land-based sources, vessels and aircraft.

The Tribunal considered – in light of the science – that states had stringent due diligence obligations under the Convention on the Law of the Sea, whereby adequate vigilance had to be applied given the high risk of harm to the marine environment in the current circumstances. It placed particular emphasis on the obligations of mitigation, including in the transboundary context, entailing not just obligations of cooperation but individual obligations by states whose inobservance would give rise to state responsibility. The Tribunal stated that the obligation of states to prevent harm includes the obligation to consider the cumulative impacts of projects (considered in Environmental Impact Assessments, both in the exclusive economic zone and high seas) prior to their approval, and developed useful guidance in relation to the obligations of states with regard to the protection of rare ecosystems and threatened species. I was ecstatic.

Sadly, Professor Boyle had passed away some time before the proceedings came to an end. I thought of him in that moment. It had been nearly six years since our trip to Katowice. In my head, I told him 'we won'. Although this had not been a contentious case, but rather Advisory proceedings, I thought it was in order to feel this way. Never before had I felt so happy to be a barrister, a specialist in international law. To me, the Advisory Opinion of the International Tribunal for the Law of the Sea was just the beginning of a new era.

I had to do something with my euphoria. That day, I looked up to the sky at night, from my balcony, and started dancing, dancing to the song *Yellow* by Coldplay (a song I associate with my work on those proceedings, as I heard it so many times while drafting) at full volume. The sky was clear that night, and it may have been my emotion, but I felt the stars looked brighter than ever.

Epilogue

My work is not done here. In fact, it is just starting. This is the beginning.

In 2021, I attended COP26 in Glasgow, but no decisive action came from that Conference of Parties. The world elites are on a 'productive denial': there is much rustling around, as if something were actively happening, but nothing happens after all. Everybody goes back to business as usual and rushes to catch a plane that leaves behind an enormous amount of waste. It is a psychological defence mechanism that Freud called '*die Verleugnung*': a person is faced with a fact that is too uncomfortable to accept and so refuses its seriousness, despite what may be overwhelming evidence. Slavoj Žižek refers to this as 'the mechanism of disavowal': you admit it but in an isolated forum; you fail to draw the consequences of it: 'I know very well, but nonetheless I don't believe, I ignore it.'[1]

It is clear to me that the changes have to come from the bottom upwards. Legal processes and litigation are part of that phenomenon.

A few weeks after the COP26 summit, back in London, I received a message from Copiapó, Chile. A Colla woman, a traditional indigenous authority, whom I met in Glasgow, got in touch to instruct me to prepare draft Articles on the Rights of Nature for the new Constitution of Chile. The Collas live in the Atacama region of northern Chile and face serious water scarcity. The exploitation of lithium by foreign companies, such as BHP, is causing serious deterioration to the aquifer resources in the area.

Paradoxically, this deterioration of permeable rock, which can contain or transmit groundwater in Atacama, is due to the extraction of a mineral to sustain a 'green economy'. Lithium is a critical component in the manufacture of batteries for electric vehicles. The paradox is not an isolated one. In Glasgow, I had already learned that Mexican indigenous peoples, the Zapotec of the Isthmus of Tehuantepec, were fighting against a proliferation of wind farms in their territories. The Isthmus sits in the migratory path of many bird species and the wind farms are already causing biodiversity loss and ecosystem disruption as a direct result of the noise and vibration of the turbines. The controversy surrounding deep-sea mining – the appetite to mine the deep sea in order to find rare minerals like lithium, used in new energy, despite the massive environmental risks, as seen in Chapter Ten – is another example to watch.

These cases make me realise that no green economy can be built on a model that takes no account of Nature. We must protect biodiversity and the world's ecosystems.

Motivated by the changes in Chile, I got down to work. The 2022 New Chilean Constitution proposal was the first in the world to be drafted in the context of the climate break down, and was to replace the one left by the Pinochet dictatorship. I appeared before the Constitutional Convention, presenting the draft articles proposal. The Rights of Nature provisions were adopted in a historic development. However, the Draft Constitution was not approved in a referendum that took place in Chile in September 2022. A total of 62 per cent of Chileans rejected the adoption of the New Chilean Constitution. Commentators observed that this was more of a 'punishment vote' to the administration, because of inflation and the sensation that 'everything is going wrong' rather than against the text itself.[2]

Despite the setback, I am persuaded that constitutional processes are essential for the protection of the natural world. I was further reassured of the importance of the rule of law to ensure respect for

Nature when I read, months later, that Gabriel Boric's new administration in Chile had initiated a lawsuit against BHP and Albermarte in Chilean courts for environmental damage caused by their operations in the Atacama, a demonstration that states can enforce their own environmental laws and regulations.

Legal action in environmental matters is succeeding in other parts of the world as well. Early in 2022, an indigenous community in Oaxaca won a five-year battle against the construction of a wind farm in the Isthmus of the Tehuantepec region in Mexico. A government agency cancelled contracts that would have allowed a French firm (Électricité de France), a leading producer of electricity worldwide, to obtain electricity from windfarms in the Zapotec territory, in Union Hidalgo, on environmental grounds.

As I am writing this epilogue, in December 2023, my mind is focused on a 1,443 km crude oil pipeline project that would transport crude oil extracted from oil fields in the Lake Albert basin in Uganda to the coast of Tanzania. The pipeline route is to go through some of the most ecologically sensitive areas in the world. If constructed, it would be the longest heated crude-oil pipeline in the world and it would put at risk of oil spills international protected wetlands, lakes on which thousands of people rely for water, endemic species, biodiversity reserves and over 118,000 people in one of the world's most seismically active regions. The massive project would potentially displace thousands, as its route goes through 231 villages in Tanzania and 178 in Uganda. I was instructed to draft a challenge against the insurer of the project, which, if successful, could stop the project from going ahead. If so, it would become the first international case in which an insurer is held to be acting contrary to the law when supporting a project that entails environmental degradation. Later, in 2024, I found myself advising a Tanzanian non-governmental organisation

(NGO) challenging the go-ahead of this project in the East African Court of Justice.

'But who are we, to tell a developing country, not to extract its natural resources?' I was asked by a fellow barrister, in his plummy English accent. The thing is, though, in this case, the company proposing to extract the oil is French and the capital is all foreign. The 118,000 people affected by the project will be driven into more poverty while the biodiversity of their country disappears. Would you call that development? As Edward O. Wilson would have put it, 'tens of thousands of species had been scraped away as by a giant hand and will not be seen in that place for generations if ever. The action can be defended (with difficulty) on economic grounds, but it is like burning a Renaissance painting to cook dinner.'[3] And, I would add, 'dinner for whom?'

While working on my submissions on that case, I feel that there is no more important work to do at this moment than saving the last bits of wilderness left in the world and the preservation of organic diversity. I also know that the law is a central tool to serve this purpose. Science has taught us that 'we are human in good part because of the particular way we affiliate with other organisms'[4] and that 'the destruction of the natural world in which the brain was assembled over millions of years is a risky step'.[5] I seek the light in the naturalists' writings. I am persuaded that giving legal protection to essential ecosystems on the surface of the Earth (including the oceans)[6] is crucial for survival, and that we should listen to science.

Some states are already doing so. In April 2022, Niue, a small island nation in the South Pacific Ocean, designated its entire exclusive economic zone – an area about the size of Vietnam – as a multiple-use marine park called Niue Nukutuluea, and 40 per cent of the park is a no-take marine protected area. Any such 'protection', however, has to be respectful of the fact that most such areas in need of protection are not *terra nullius* but the ancestral land of indigenous peoples. The idea of protection goes rather to the notion of intangibility (protection from logging, from mining, from oil/gas exploitation). The Amazon, for example, is inhabited by many tribal groups that are known to live in harmony with Nature. This is their ancestral land.

The law, and ensued action, has to make more strides.

I am presented with a paradox that looks like this: somewhere in the Peruvian Amazon an oil spill seeping into the land and rivers takes place. Yet, it is not possible to sue the company responsible for the spill (a company incorporated in England) in English courts, because no human being affected by the spill can be identified. Under English law, a tort occurs only when someone commits a wrong against another person. But what if the injury is to Nature? Could it be possible to argue that, in the absence of a human being, there was no harm? If there is a gap in the law, isn't it time to correct such failure?

A news item prompted further reflections. I learned, in November 2022, that the last man from an uncontacted tribe in the Amazon, who had lived in voluntary isolation, a Tanaru man, had died in his ancestral land. With him, the DNA of the Tanaru, together with their language and culture as a tribe, was erased from the world. Today, there is discussion as to whether the protection status of the Tanaru reserve, Rondônia, now with no uncontacted people, should be revoked. 'The land was protected because of the indigenous man's presence', is the central argument. He was eventually buried in the Tanaru reserve. I say: this alone, the respect for his remains, should keep Rondônia protected, as a Tanaru ancestral land, in accordance with international human rights law. However, shouldn't the Tanaru reserve make sense as a concept also because of the protection afforded to the living world other than human life?

Until we have given legal recognition to such a living world, we shall continue to permit the burning of irreplaceable parts of the world that hold the key to the future, those Renaissance-like pieces of art (in the analogy of Edward O. Wilson) that provoke our wonderment because they contain both our own origin and our future. After all, as theoretical physicist Stephen Hawking illuminated, human beings are mere collections of fundamental particles of nature.

In the small *maloca* in which the Tanaru man lived, now lies his place of burial. To my mind, this marks the site of a modern genocide, that of an Amazonian tribe and its culture. Several indigenous movements have called for the Tanaru reserve to be protected as a memorial to indigenous genocide and maintained as an environmental conservation area.

I close this book with hope. I have been woken up by a chatty, garrulous, song thrush early in the morning. Intriguingly, bird song soothes me. Add to that a reassuring woodpecker's drumming in the background, as if saying 'all shall be well, and all shall be well, and all manner of thing shall be well' like Julian of Norwich was told in her visions.[7] It is not only the three international courts' respective advisory opinion mechanisms that have now been seized to deliver Opinions on climate change, as mentioned earlier, but also ordinary people fighting to protect nature around the world.

Are you able to see 'all that is made', a miniature reflection of the universe, in a hazel nut?[8]. Have you conceived that we may be, like the hazel nut, mere fractals of the natural world, of the universe?

Building on the growing momentum gained by the 'rights of nature' approach, a community in Lewes, East Sussex, has started an unprecedent process in England. On 20 February 2023, Lewes council passed a motion that aims to recognise the River Ouse's rights to protection amid growing concerns over pollution of waterways. This marked the first step towards developing a Rights of River Ouse Charter to secure its health and its right to be pollution free. The charter, which is to be drafted over the next two years, aims to enshrine substantive rights to the river, 'redefine human-river interaction', and promote a sustainable river system.

I have been instructed by the Environmental Law Foundation to assist in drafting the charter, the first of its kind in England, and to advise on its operationalisation, working with the local community who have organised as a group called Love Our Ouse. I met Emma Montlake, a lawyer at the Environmental Law Foundation and a citizen of Lewes, and Councillor Matthew Bird on my first visit. It is encouraging to see a community taking such initiative, one likely to inspire other communities in England and around the world.

Following my earlier visit, one Sunday I carried out some research. I started becoming acquainted with the River Ouse. Firstly, I realised that it is not a single body of water. It is a *network* of water bodies. Their tributaries, including chalk streams, must be protected as well if the health of the river is to be restored and preserved. I also discovered that the River Ouse goes through and supports life in an area possessing an array of sites designated as being of international, national and local importance for biodiversity. These include one Special Protected Area (SPA); three Special Areas of Conservation (SAC); two National Natural Reserves; and twenty-four Sites of Special Scientific Interest. Many of these sites support important wetlands habitats and species sensitive to changes in water level, flow and quality.

The fauna within the River Ouse catchment area also inspires me. My research felt like going on safari. The names of the species

I discovered sound like creatures out of a ballet: 'swan mussels', 'pea mussels', 'freshwater shrimp'. The freshwater shrimp is vulnerable in the Red Data Book and fully protected in the UK under the Wildlife and Countryside Act 1981.[9] The freshwater shrimp is very sensitive to oxygen levels, so it is an indicator of water pollution levels. If freshwater shrimps are found in the river, it means that the river is clean. Therefore, enforcing the protection of the freshwater shrimp is a way of ensuring the non-pollution of the river more generally. Another species I found is the pea mussel, which is an endangered species and protected by the Natural Environment and Rural Communities Act 2006, Section 41. It is also an indicator of purity of water. There are also swan mussels, whose shell colour is often greenish or brownish. This is an important species in the aquatic ecosystem because it cleanses the water of algae, bacteria and other particulate matter.

I am starting to love the Ouse. I feel like the Mole in Kenneth Grahame's *The Wind in the Willows*, 'bewitched, entranced, fascinated'; spellbound.

I crossed the Sargasso Sea and found myself, the next day, looking up at the sky, swimming in the ocean at 5 a.m.

Life had brought me, on 19 April 2024, to Bridgetown, Barbados, for another historic case. I had just arrived in Barbados to participate in oral proceedings before the Inter-American Court of Human Rights, which was extraordinarily sitting there in a *Climate Emergency and Human Rights* Advisory Opinion case. The courtroom was set up in the University of West Indies, and the participation of several state delegations, which included Barbados, attracted a lot of security personnel.

On 9 January 2023, Chile and Colombia triggered the advisory jurisdiction of the Court, asking questions on what the obligations for states were to address the climate emergency within the framework of human rights law. Another international development which the first one, the *Torres Strait*, had paved the way for. These entailed formal Court proceedings whose first sitting took place from 23 to 25 April 2024. The response the Court would give in the Opinion would constitute an authoritative pronouncement on the obligations of states in the Americas. Around 265 written submissions were filed by states, intergovernmental organisations and other parties. What was being fought at the hearing in Barbados would define the future of the climate in the American region.

On my first morning in Barbados, floating facing the sky, I closed my eyes, trying to make contact with the sea. I wanted to sense it. The smell (albeit more subtle) took me right back to my childhood. To the smell of Pucusana, the first place I swam in the sea. The smell of the Pacific Ocean on the coast of Peru, in my memory, was wild, ferociously so. I've never again encountered a sea that had the intensity of that stretch of the Pacific, with its sea birds and the strong smell of seaweed. By contrast, in Barbados, on the west coast where the water is calm, the sea has a transparent turquoise colour and is mild. In that water I had a magical encounter. I was swimming in the sea when a green sea turtle came alongside me. The turtle was swimming so effortlessly, almost floating, weightless, as in a dream. It was going down to the seabed, to eat something off the sand before going up to catch some oxygen from time to time. I was transfixed and, for a while, we swam together. The dream-like memory I have of our encounter, watching it moving below me, felt like it happened in slow motion.

That evening, a chorus of whistling frogs, ever present in the dense vegetation of the grounds where I was staying, also felt magical.

In the buddhist tradition, the probability of a life-form coming into existence in the shape of a human being is like a turtle surfacing from the bottom of the sea and coming up through a ring the size of its body. We are rare. Shouldn't we make something wonderful of our time on Earth? Aren't we responsible for those other life forms who are affected by our acts?

I appeared before the Court on 23 April 2024. A lot is pending not only before this Court, but also before the International Court of Justice, which is now grappling equally with the topic of climate change and global warming. But, at this point, I want to celebrate the power of the law and of ordinary people, in forcing the law to address environmental harms.

This is happening all around the world. Another way of living *is* possible.

Notes

Introduction

1 Julian Hoffman, *Irreplaceable* (Penguin, 2020), p. 15.
2 Charles Darwin, *A Naturalist's Voyage Round the World* (John Murray, 1901), p. 369.
3 Simone Weil, *Gravity and Grace* (Ark Paperbacks, 1987), p. 116.
4 Andrea Wulf, *The Invention of Nature: The Adventures of Alexander von Humboldt, the Lost Hero of Science*, (John Murray, 2016), p. 88.
5 Simone Weil, *Gravity and Grace*, op. cit., p. 141.

1. The Wayúu and the Largest Open-Pit Coal Mine in Latin America

1 Inter-American Commission on Human Rights, Resolution 60/2015 (Provisional Measures No 51/15), *Asunto niñas, niños y adolescentes de las comunidades de Uribía, Manaure, Riohacha y Maicao del pueblo Wayúu, asentados en el departmento de la Guajira, respecto de Colombia*, p. 1 (11 December 2015), https://perma.cc/3DUZ-RMPL.
2 Adrian Alsema, 'La Guajira's desert and its legendary Wayuu people', *Colombian Reports*, 25 September 2019.
3 Rúa Bustamante et al, 'Rol de las mujeres de las comunidades indígenas wayuu en la Guajira (Colombia) en torno a la actividad ganadera', in Marquez Dominguez and Llamas Chavez (eds), *Hélices y anclas para el desarrollo local* (2019).
4 Organización Binacional de Mujeres Wayúu, 'La Espiritualidad Wayúu', https://organizacionbinacionalmujereswayuu.es.tl/LA-ESPIRITUALIDAD.htm.
5 Inter-American Commission on Human Rights, Resolution 60/2015 (Provisional Measures No 51/15), p. 2.
6 Colombian Constitutional Court, T-614 of 2019, relating to the right to health of indigenous communities (*Acción De Tutela Para Proteger Derecho a la Salud y Ambiente Sano de Comunidad Indígena*

Frente a Actividades Extractivas de Carbon-Procedencia; hereinafter Colombian Constitutional Court, T-614), https://perma.cc/W7MS-8LDK, para. 9.10 [Author's translation].

7 'Minería a gran escala y derechos humanos: lo que el des-arroyo trajo a la Guajira', Revista *Noche y Niebla*, No. 61 (Enero-Junio 2020), CINEP, p. 65 [Author's translation].

8 CAJAR, 'Diez Verdades sobre Carbones del Cerrejón – Report', (2019), p. 7, https://perma.cc/529Z-GBG2. This includes the municipalities of Fonseca, Barrancas, Hato Nuevo, Albania, Uribía and Maicao in the department of La Guajira.

9 Ibid, p. 6.

10 Colombian Constitutional Court, T-614, para. 9.8 [Author's translation].

11 Fueyo Editores, 'Cerrejón, la explotación de carbón más grande de todo Sudamérica' (Artículo técnico), 24 March 2015.

12 Colombian Constitutional Court, SU698/17 of 2017, relating to the *Arroyo Bruno* case [Author's translation].

13 Testimony of a member of the communities of Roche, Las Casitas and Tabaco in 'Las Huellas del Cerrejón', Documentary, https://www.youtube.com/watch?v=ryssy7pJhJI [Author's translation].

14 On the relative dangers of PM 10 and PM 2.5 in causing lung cancer, see Ole Raaschou-Nielsen et al, 'Air pollution and lung cancer incidence in 17 European cohorts: prospective analyses from the European Study of Cohorts for Air Pollution Effects (ESCAPE)', *The Lancet* (2013), 14(9), 813–822.

15 CAJAR, 'Diez Verdades sobre Carbones del Cerrejón', op. cit., p. 18.

16 Monitoring of PM 2.5 at Cerrejón only began following a Colombian domestic regulation introduced in 2018. See Resolution No. 2254, by which the ambient air quality standard is adopted and other provisions are dictated, available at https://perma.cc/2PEM-NR6N; see also Gabriel Bustos, 'New Air-Quality Regulation' (Cerrejón, 1 March 2018), https://perma.cc/DC6S-BHDB.

17 See the findings in Laura Rodríguez A. (ed), *Carbón Tóxico: Daños y riesgos a la salud de trabajadores mineros y población expuesta. Evidencias científicas para Colombia*, Fundación Rosa Luxemburg, October

2018 https://www.rosalux.org.ec/pdfs/FRL_Carbon_toxico_WEB_compressed.pdf .

18 'Colombia – The Curse of Coal' (2017), available at https://www.youtube.com/watch?v=t1xLZWp2eBc&ab_channel=DWDocumentary [last accessed 28 October 2020].

19 Hilda Lloréns and Ruth Santiago, 'Coal's Open Wounds / Las Heridas Abiertas del Carbón' (NACLA, 28 September 2018), https://perma.cc/6XYB-WT2V.

20 Constitutional Court of Colombia, Judgment T-614.

21 The Colombian Constitutional Court ruled that, as an urgent transitional measure, Carbones del Cerrejón must control its mean emissions (calculated by month) to a maximum of 20 µg/m3 for PM 10 and 10 µg/m3 for PM 2.5. The measure was to remain in force until agreement on particulate matter limits was to be reached between Carbones del Cerrejón and the affected communities. Constitutional Court of Colombia, T-614, Order 3 [Author's translation].

22 Alcides R. Daza-Daza, Nelson Rodriguez-Valencia and Alexis Carabali-Angola, 'El Recurso Agua en las Comunidades Indígenas Wayúu de la Guajira Colombiana. Parte 1: Una Mirada desde los Saberes y Prácticas Ancestrales', 29(6), pp. 13–24, *Información Tecnológica*, La Serena, Dic. (2018), 29(6), p. 14. http://dx.doi.org/10.4067/S0718-07642018000600013

23 Ibid., citing Instituto Colombiano de Cultura Hispánica, Geografía Humana de Colombia, 2014 [Author's translation].

24 'Minería a gran escala y derechos humanos: lo que el des-arroyo trajo a la Guajira', op. cit., p. 67 [Author's translation].

25 Golda Amanda Fuentes, Jesús Olivero Verbel, Juan Carlos Valdelamar Villegas, Daniel Armando Campos, Alan Philippe, 'Si el río suena, piedras lleva', (Sobre los Derechos al Agua y a un ambiente sano en la zona minera de la Guajira), Indepaż, 2018, p. 23, referring to Corpoguajira, Action Plan 2016–2019 (Plan de acción 2016–2019 prosperidad, paz y sostenibilidad), p. 13. http://www.indepaz.org.co/wp-content/uploads/2019/02/Si-el-rio-suena-piedras-lleva-Indepaz-2019.pdf

26 Marcela Zuluaga Contreras, 'El único río en el desierto', *El Turbión*, 10 November 2021.

27 Lydia James, 'Dangerous levels of mercury found in river in Colombian region of La Guajira', London Mining Network, 25 November 2019, https://perma.cc/VXG8-676M.
28 Cerrejón, 'Cerrejón's Sustainability Report 2019' (2019), p. 49.
29 Ibid: 'Rainwater runoff and coal seam water provide 89 per cent of the water used in our processes, primarily for reducing particulate matter emissions. The remaining 11 per cent is freshwater withdrawn from the mid-valley of the Ranchería River and its alluvial aquifer.' Freshwater extraction is shown to total 1,241 megalitres.
30 CAJAR, 'Diez Verdades sobre Carbones del Cerrejón', op. cit., p. 9.
31 Informe de Sostenibilidad de Cerrejón, 2018, p. 82, https://www.colectivodeabogados.org/?10-verdades-sobre-Carbones-del-Cerrejon.
32 Comisión Interamericana de Derechos Humanos Resolución 60/2015 (Medidas Cautelares No 51/15), op. cit., p. 1 [Author's translation].
33 CENSAT Agua Viva, 'La desviación del arroyo Bruno: entre el desarrollo minero y la sequía' (2015), p. 9, https://perma.cc/QYB7-LLL7.
34 'Cerrejón's Sustainability Report 2019', op. cit., p. 49.
35 Daniel Armando Campos and Allan Philippe, 'Monitoring and assessment of polluting metals in the southeastern mining-impacted region of La Guajira, Colombia' (University of Koblenz Landau, 2017), p. 35.
36 Ibid.
37 Informe de Resultados de Laboratorio, 9 July 2019, https://perma.cc/53N6-BX2F. See also Lydia James, 'Dangerous levels of mercury found in river in Colombian region of La Guajira', op. cit. The lab results showed that the mercury level was 0.0749 mg/L. The World Health Organization sets the mercury guidelines at 0.006 mg/L, while the Colombian Government's Resolution 2115 (22 June 2007) set the recommended limit at 0.001 mg/L, https://perma.cc/K5GX-H7EQ.
38 World Health Organization, 'Zinc in Drinking-water', WHO/SDE/WSH/03.04/17 (2003), p. 3.
39 World Health Organization, 'Cadmium in Drinking-water', WHO/SDE/WSH/03.04/80/Rev/1 (2011), p. 5.

40 World Health Organization, 'Barium in Drinking-water', WHO/FWC/WSH/16.48 (2016), p. 9.
41 World Health Organization, 'Manganese in Drinking-Water', WHO/SDE/WSH/03.04/104/Rev/1 (2011), p. 11.
42 World Health Organization, 'Lead poisoning and health', 23 August 2019, <https://perma.cc/8JQT-HMHM>; World Health Organization, 'Iron in Drinking-water', WHO/SDE/WSH/03.04/08 (2003), p. 3.
43 World Health Organization, 'Mercury and health', 21 March 2017, https://perma.cc/YJ2H-PZEJ.
44 Constitutional Court of Colombia, T-614, p. 123, para. 9.7 [Author's translation].
45 Ibid, para. 11.14 [Author's translation].
46 'Cerrejón's Sustainability Report 2019', op. cit., p. 51.
47 Ibid, pp. 11–13, 25, 40, 43, 45.
48 See letters from ABColombia to the Corporación Autónoma Regional de la Guajira and to Carbones del Cerrejón, 31 July 2018, https://perma.cc/8FHB-ELSR.
49 The plight of Luz Angela Uriana Epiayú was featured in a documentary by DW, *Colombia: The curse of coal* (2017), which can be watched in the following link: https://www.youtube.com/watch?v=t1xLZWp2eBc.
50 ICESCR Committee, General Comment No. 15, para. 16, d. https://www.refworld.org/pdfid/4538838d11.pdf.
51 Ibid, at para. 7.
52 Inter-American Commission on Human Rights, Advisory Opinion OC-23/17, 15 November 2017 (Advisory Opinion 23) para. 59.
53 Ibid, para. 47.
54 Inter-American Court of Human Rights, *Case of the Indigenous Communities of the Lhaka Honhat (Our Land) Association v Argentina*, Judgment of 6 February 2020 (Merits, Reparations and Costs) § 209.
55 'Minería a gran escala y derechos humanos: lo que el des-arroyo trajo a la Guajira', op. cit., p. 79.
56 CAJAR, 'Diez Verdades sobre Carbones del Cerrejón', op. cit., p. 9.
57 Mauricio Ramírez, 'La red hídrica afectada por la explotación minera de la Guajira', 30 January 2018, https://perma.cc/RXH7-YG33.

58 Tathiana Montaña, 'La desviación del Río Ranchería: algunos elementos a consideración' (Notas visita Riohacha 16–18 October 2012, Indepaz), https://perma.cc/U8KA-QQJ7.
59 Ibid, p. 3. Author's translation.
60 Ibid, p. 2. Author's translation.
61 Lydia James, 'British multinational disobeys Colombia court by diverting water source', London Mining Network, 8 July 2019, https://perma.cc/LZE9-J8JT.
62 Ibid.
63 Richard Solly, 'Saving the river: The struggle for Colombia's Arroyo Bruno', London Mining Network, 20 July 2019, https://perma.cc/CSX7-8G4B; Colectivo de Abogados José Alvear Restrepo (CAJAR), 'Carbones del Cerrejón mantendrá taponado el Arroyo Bruno desconociendo fallo de la Corte Constitucional', 12 June 2019, https://perma.cc/BD9F-L963.
64 See, for example, Inter-American Court of Human Rights, *Case of the Saramaka People v Suriname*, Preliminary Objections, Merits, Reparations and Costs, Judgment of 28 November 2007 at E.2 (a), para. 133.
65 Ibid, para. 130.
66 Inter-American Court of Human Rights, *Case of the Saramaka People v Suriname*, Interpretation of the Judgment of Preliminary Objections, Merits, Reparations and Costs, Judgment of 12 August 2008, para. 40.
67 Ibid, para. 41.
68 Inter-American Commission on Human Rights (IACHR), Indigenous and Tribal Peoples' Rights over their Ancestral Lands and Natural Resources: Norms and jurisprudence of the Inter-American Human Rights System, OEA/Ser.L/V/II, 30 December 2009, para. 233; Inter-American Court of Human Rights, *Case of the Saramaka People v Suriname*, Interpretation of the Judgment of Preliminary Objections, Merits, Reparations and Costs, Judgment of 12 August 2008, para. 37.
69 Inter-American Court of Human Rights, *Case of the Saramaka People v Suriname*, Interpretation of the Judgment of Preliminary Objections, Merits, Reparations and Costs, Judgment of 12 August 2008, para. 37.

70 Colombian Constitutional Court, T-528 of 1992. The Court relied on resolution 02122, issued by the Ministry of Health (12 February 1992), which had identified the area surrounding the Carbones del Cerrejón mine as 'uninhabitable' and dangerous to human and animal life, and plant agriculture. The Court ordered the authorities to 'ensure the preservation of the quality of life and a health environment.'
71 Judgments against Carbones del Cerrejón include the following: Colombian Supreme Court (Corte Supreme de Justicia) 13/0912, 0014-01 of 7 May 2002; Criminal Cassation Chamber of the Supreme Court, sentence of 13 September 2012; Colombian Constitutional Court, T-256 of 2015; Colombian Constitutional Court, T-704 of 2016; Barrancas Court (Juzgado Promiscuo de Barrancas) 2015-00473 of 26 February 2016; Administrative Supreme Court of Colombia (Consejo de Estado) 2016-00079-01 of 13 October 2016; Colombian Constitutional Court, SU-698 of 2017; Colombian Constitutional Court, T-329 of 2017.
72 Richard Solly, 'Legal action against Cerrejón Coal's environmental licence', London Mining Network, 27 February 2019, https://perma.cc/UC4Q-ZSQP.
73 Colombian Constitutional Court, SU-698 of 2017.
74 Ibid.
75 Ibid.
76 [Author's translation].
77 William Avilés, 'The Wayúu tragedy: death, water and the imperatives of global capitalism', *Third World Quarterly* (2019), 40(9), pp, 1750–1766 at p. 1750.
78 Inter-American Commission on Human Rights, Resolution 60/2015.
79 Ibid, para. 1. See also CAJAR, 'Diez Verdades sobre Carbones del Cerrejón', op. cit., p. 15; Defensoría del Pueblo de Colombia (Ombudsman's Office of Colombia), 'Crisis humanitaria en La Guajira 2014', June 2014, https://perma.cc/JWH9-46AJ.
80 CAJAR, 'Diez Verdades sobre Carbones del Cerrejón', op. cit., p. 17.
81 'Guáimaro, un guardián ambiental nacido de la entraña de la Tierra', *El País*, 22 March 2018, https://perma.cc/7YMM-H5GC.

82 'Guáimaro, guardián ambiental nacido de la entraña de la Tierra', *El Nuevo Siglo*, 19 March 2018, https://perma.cc/K3G7-C493.

83 Ambiente, 'Guáimaro, el árbol que renace para luchar contra la desnutrición en Colombia', *El Espectador*, 20 March 2018, https://perma.cc/XK9M-RA7V.

84 See, for example, Sentence SU689/17 for the many references to the Guáimaro being affected by the diversion of the Arroyo Bruno; https://perma.cc/SPN3-V3KE.

85 Constitutional Court of Colombia, T-614, p. 123, para. 9.7 [Author's translation].

86 The Wayúu, Afro-Colombian and campesinos communities have traditionally used over 170 types of plants in their medicines. The growth of these plants has been impeded by the operation of the mine, 'Minería a gran escala y derechos humanos', op. cit., p. 84.

87 For a recognition of the intangible cultural heritage of the Wayúu people, see UNESCO, 'Decision of the Intergovernmental Committee: 5.COM 6.9' (2010), https://ich.unesco.org/en/RL/wayuu-normative-system-applied-by-the-putchipuui-palabrero-00435.

88 Katrin Blanta, 'Interdependency and Interference: The Wayuu's Normative System and State-based Conflict Resolution in Colombia', Berghof Foundation (2016), p. 12, https://perma.cc/CYV7-ANTU.

89 Statement by Rogelio Ustate Arrogoces; see Hilda Lloréns and Ruth Santiago, 'Coal's Open Wounds / Las Heridas Abiertas del Carbón' (NACLA, 28 September 2018), https://perma.cc/6XYB-WT2V.

90 'Minería a gran escala y derechos humanos', op. cit., p. 75. In particular, the communities of Palmarito, Caracolí, Espinal and Tabaco have been affected.

91 [Author's translation].

92 Christian Cwik, 'Displaced Minorities: The Wayuu and Miskito People', in Steven Ratuva (ed.), *The Palgrave Handbook of Ethnicity* (Palgrave, 2019), pp. 1593–1609. https://perma.cc/LWN3-F93K.

93 They examined Ireland 5th to 9th combined reports.

94 CERD/C/IRL/CO/5-9, 12 December 2019. The full report can be found at: https://tbinternet.ohchr.org/Treaties/CERD/Shared%20 Documents /IRL/INT_CERD_COC_IRL_40806_E.pdf.

95 See the German Parliamentary Enquiry 'Coal Imports on Colombia' (*Steinkohleimporte aus Kolumbien*), 28 January 2019. In 2015 Germany imported 6846 t, in 2016 8054 t and in 2017 4670 t of coal from Colombia; https://dip21.bundestag.de/dip21/btd/19/074/1907405.pdf. See also the report 'Steinkohle aus Kolumbien: Deutschlands Kohlepolitik Im Kontext der Energiewende', produced by a group of scientists and activists in Aachen, stating that Colombia exports 95 per cent of coal produced, with two-thirds going to Europe, and, in Europe, 'Germany is by far the biggest buyer'; see p. 10, available at: https://resilienz-aachen.de/wp-content/uploads/2019/11/Dominik-Johannes-Gasser-Steinkohle.pdf.

96 'Steinkohle aus Kolumbien: Deutschlands Kohlepolitik Im Kontext der Energiewende' , ibid, p.10.

97 See https://www.banktrack.org/project/cerrejon_coal_mine.

98 London Mining Network, 30 April 2016, 'HSBC under attack for coal financing, human rights abuse, dodgy deals and more', https://londonminingnetwork.org/2016/04/hsbc-under-attack-for-coal-financing-human-rights-abuse-dodgy-deals-and-more/.

99 Global Justice Now, 'Mapping Dirty Finance', May 2013. See also Global Justice Now, 'The Cerrejón Mine, Coal exploitation in Colombia', May 2013.

100 See https://www.banktrack.org/download/158e31a/cerrejon_media_briefing.pdf.

101 A/74/480. Report of the Special Rapporteur on the implications for human rights of the environmentally sound management and disposal of hazardous substances and wastes, 7 October 2019. https://docs.un.org/en/A/74/480.

102 A/HRC/45//12 §5. Duty to prevent exposure to the COVID-19 virus, Report of the Special Rapporteur on the implications for human rights of the environmentally sound management and disposal of hazardous substances and wastes, 13 October 2020, § 5. https://docs.un.org/en/A/HRC/45/12.

103 Ben Butler, 'BHP commits to selling its thermal coalmines within two years', *Guardian*, 18 August 2020.

104 Ibid.

105 UN, 'David R. Boyd, Special Rapporteur on human rights and the environment', <https://perma.cc/959U-F926>.
106 Special Rapporteur David Boydd's video recording, https://www.youtube.com/watch?v=ffWTT9Q69g8.
107 UN, 'UN expert calls for halt to mining at controversial Colombia site', 28 September 2020, https://perma.cc/5UT9-JKBG.
108 Neil Hume, 'Big miners face new front in Colombian human rights battle', *Financial Times*, 19 January 2021, https://www.ft.com/content/f63adfa4-1b63-4e9c-809e-138815d9ee50.

2. Myanmar: Southeast Asia's Last Free-flowing River

1 Hemi Kim, 'Why the term "sentient" is so complicated', *Sentient Media*, 8 September 2021.
2 Reported in a 1586 report by a Spanish Corregidor. See, Ulloa Mogollón, Juan de (1965 [1586]), 'Relación de la Provincia de los Collaguas para la descripción de las Indias que su majestad manda hacer', in Marcos Jiménez de la Espada (ed.), *Relaciones Geográficas de Indias*, vol. I, pp 326-333. Biblioteca de Autores Españoles: Perú 183. Atlas, Madrid (1965 [1586]) .
3 Paul H. Gelles, 'Caballeritos and Maiz Cabanita: Colonial Categories and Andean Ethnicity in the Quincentennial Year', KAS Papers, Berkeley University, p. 19.
4 Eric Hilaire, 'China's Plan to build hydro dams over Salween river in pictures', *Guardian*, 29 January 2013, https://www.theguardian.com/environment/gallery/2013/jan/29/china-salween-river-hydro-dams.
5 International Rivers, 'The Salween River Basin Fact Sheet', 24 May 2012, https://archive.internationalrivers.org/resources/the-salween-river-basin-fact-sheet-7481.
6 The remarkable biological diversity of the area in China through which the Salween flows, secured UNESCO World Heritage Site status in 2003.
7 International Rivers, 'The Salween', https://www.internationalrivers.org/where-we-work/asia/salween/.
8 International Rivers, 'Salween', https://www.internationalrivers.org/where-we-work/asia/salween/.

9 Demelza Stokes, '"My Spirit is there": Life in the shadow of the Mong Ton dam', Mongabay, 7 December 2016, https://news.mongabay.com/2016/12/my-spirit-is-there-life-in-the-shadow-of-the-mong-ton-dam/.
10 Edward O. Wilson, *Half-Earth* (Liveright, 2016), p. 155.
11 Ibid, p. 106.
12 Darlene Lee, 'Do Dams Violate a River's River to Flow?', Earth Law Center, 17 October 2017.
13 Ibid.
14 Ibid.
15 'The largest dam demolition in history is approved for a Western river', *NPR*, 17 November 2022.
16 Ibid.
17 Demelza Stokes, 'My Spirit is there', op. cit.
18 International Rivers, 'The Salween River Basin Fact Sheet', op. cit.
19 Collectively, the dams were estimated to produce more than 15,000 MW of electrical power, which far exceeds contemporary domestic energy needs in Myanmar.
20 Kyaw Phone Kyaw, 'Thanlwin dam projects "unjust": civil society', *Myanmar Times*, 9 July 2015, https://www.mmtimes.com/national-news/15416-thanlwin-dam-projects-unjust-civil-society.html.
21 The Third Pole, 'China's Salween plans in limbo in post-coup Myanmar', 8 June 2021, https://www.thethirdpole.net/en/energy/chinas-salween-plans-in-limbo-in-post-coup-myanmar/.
22 Pianporn Deetes, 'Today is a Day of Action for Rivers', *Bangok Post*, https://www.bangkokpost.com/opinion/opinion/2278643/today-is-a-day-of-action-for-rivers.
23 Andrew Paul, Saw Sha Bwe Moo, and Robin Roth, 'Water and fish conservation by Karen communities: An indigenous Relational Approach', International Institute for Asian Studies, Newsletter, 94, Spring 2023.
24 Ibid.
25 https://www.internationalrivers.org/take-action/international-day-of-action-for-rivers/.
26 Wilson, *Half-Earth*, op. cit., p. 58.
27 United Nations Conference on the Human Environment in Stockholm, 5–16 June 1972.

28 United Nations Conference on Environment and Development (UNCED) in Rio de Janeiro, 3–14 June 1992.

29 Wilson, *Half-Earth*, p. 187.

30 Inter-American Commission on Human Rights, Advisory Opinion OC-23/17, 15 November 2017, para. 62.

31 Ibid, para. 62.

32 *Bebb v. Law Society*, C. A. [1913 B. 305], https://www.law.cornell.edu/sites/www.law.cornell.edu/files/women-and-justice/Bebb-v-Law-Society.pdf.

33 ibid., p. 294

34 Benedict de Spinoza, *Ethics* (Penguin, 1996).

35 Steven Nadler, *Spinoza: A Life* (Cambridge University Press, 1999), p. xi.

36 As suggested by Ajay Singh Chaudary, 'God or Nature: Spinoza's Ethics, Brooklyn Institute of Social Research, https://thebrooklyninstitute.com/items/courses/new-york/god-or-nature-spinozas-ethics-4/#:~:text=Rejecting%20Descartes'%20view%2C%20Spinoza%20argued,single%20substance%3A%20God%20or%20Nature.

37 Upon being asked if he believed in God by Rabbi Herbert Goldstein of the Institutional Synagogue, New York, April 24, 1921, *Einstein: The Life and Times*, Ronald W. Clark, (Bloomsbury Reader, 1972) p. 502.

38 As suggested by Stuart Hampshire, in his introduction to Benedict de Spinoza, *Ethics*, op. cit., p. ix.

39 Jonathan Israel, *Spinoza: Theological Political Treatise* (Cambridge Texts in the History of Philosophy, 2007), p. 195.

40 Ibid, p. 196.

41 Michael Black, KC, 'Do arbitrators dream of electric parties?', Tylney Hall Dinner Speech, 18 May 2024.

42 Ibid.

43 Andrea Wulf, *The Invention of Nature: The Adventures of Alexander von Humboldt, The Lost Hero of Science*, op. cit., p. 88.

44 Constitutional Court of Ecuador, Judgment No. 218-15EP-CC, 9 July 2015.

45 Constitutional Court of Colombia, Judgment T-622-16, 10 November 2016.

46 High Court of Uttarakhand at Naintal, India, Judgment of 30 March 2017.
47 The Minganie County in the Province of Quebec, Canada, issued a Municipal Resolution recognising legal personhood and rights (including the right to flow, the right to be free from pollution, the right to regeneration) of the Magpie River, Province of Quebec, Municipality of Minganie County, Resolution no.025-21, Recognition of the legal personality and rights of the Magpie River – Mutehejau Shipu, 17 February 2021.
48 Juzgado Mixto Nautua I, Resolución No 14, 8 March 2024, Acción de amparo, granting rights to the Marañon River.
49 Ruling T-622/16 delivered by the Constitutional Court of Colombia, on 10 November 2016. https://justiciaambientalcolombia.org/2017/05/07/sentencia-rio-atrato/ [accessed 27 April 2018].
50 See https://www.parliament.nz/en/get-involved/features/innovative-bill-protects-whanganui-river-with-legal-personhood/ [accessed 28 April 2018].
51 Ibid.
52 José María Arguedas, *Los Rios Profundos*, Editorial Losada (Buenos Aires, 1958).
53 Apu means 'Lord' in Quechua. It refers to mountains, which in the Andean tradition have a spirit that is alive. That is where people found the spirits of their ancestors. https://en.wikipedia.org/wiki/Apu_(god).
54 See http://povmagazine.com/articles/view/review-daughter-of-the-lake. On the resistence of the local community to Yanacocha, see https://www.theguardian.com/environment/2016/apr/19/goldman-prize-winner-i-will-never-be-defeated-by-the-mining-companies [accessed 27 April 2018].
55 [Author's translation].
56 Article 33 [Author's translation].
57 See http://pdba.georgetown.edu/Constitutions/Ecuador/english08.html [accessed 27 April 2018].
58 Ralph Waldo Emerson, *Nature* (James Munroe and Company, 1836), p. 5.
59 Ibid, p. 11.
60 Cormac Cullinan, *Wild Law* (Chelsea Green Publishing, second edition 2011), p. 41.

61 T-622-16, §9.32 [Author's translation].
62 Ibid.
63 T-622-16, pp. 166–7.
64 Ibid, p. 182.
65 Ibid.
66 Universal Declaration of the Rights of Rivers https://static1.squarespace.com/static/55914fd1e4b01fb0b851a814/t/5c93e932ec212d197abf81bd/1553197367064/Universal+Declaration+of+the+Rights+of+Rivers_Final.pdf.
67 Earth Law Center, Universal Declaration of River Rights, https://www.earthlawcenter.org/river-rights.
68 Ibid.
69 See https://www.reuters.com/article/us-colombia-deforestation-amazon/colombias-top-court-orders-government-to-protect-amazon-forest-in-landmark-case-idUSKCN1HD21Y.
70 Judgment STC4360-2018, 5 April 2018, http://legal.legis.com.co/document?obra=jurcol&document=jurcol_c947ae53aeb447bd91e8e9a315311ac5 [accessed 25 April 2018].

3. The Dark Business of Light in the Land of Birds

1 Judith Adalgasia del Carmen Franco Sandoval, 'Monografía de Chiquimula Educación y Cultura', Tesis de Maestría, Universidad de San Carlos de Guatemala, July 2003, p. 18.
2 Nuevo Día Ch'orti' Indigenous Association (CCCND).
3 OEA/Ser.L/V/II.118 Doc, 5 rev. 1, 29 December 2003, Inter-American Commission on Human Rights' Thematic Report on Guatemala, *Justice and Social Inclusion: The Challenge of Democracy in Guatemala*, at para. 216.
4 In 2013, Rios Montt was found guilty of the genocide of indigenous peoples and condemned to eighty years in prison. Two weeks later the country's constitutional court overturned the verdict on technical grounds and ordered a retrial. In the retrial set for January 2015, Montt was deemed too unwell to attend. Ríos Montt was to face a second Genocide Trial for the Dos Erres Massacre. Special provisions were adopted in these proceedings given the aging Ríos Montt's state of health. Neither trial was completed by the time Ríos Montt died in April 2018.

5 The Agreement on Identity and Rights of indigenous Populations, which represented the historical opportunity to ending exclusion and discrimination of indigenous populations in Guatemala, among them.

6 'The census lies, it minimizes the Ch'orti' people.' ('*El censo miente, minimiza al pueblo Ch'orti*'), interview with Jeremías, representative of Nuevo Día'. The Ch'orti' spread around Camotán, Jocotán and beyond, including Olopa, Unión Zacapa and other municipalities of Zacapa.

7 '*Es de nosotros, sigue siendo de nosotros. Es nuestro patrimonio, desde nuestros ancestros. Nos lo dejaron nuestros Abuelos.*'

8 '*Que es el río Jupilingo? Es un milagro ver ese río porque 3 años que se pierde cosecha por la sequía. Es un milagro. Es el único bien natural que nos da la vida, de eso vivimos.*'

9 '*Cuando no hay agua, tenemos que ir al río.*'

10 '*Comemos de la tierra.*'

11 'The animals, cattle, cows, go to the river to drink water.' ('*Los animales, las reses, vacas, van al río a beber agua.*')

12 While the census identifies only 40 per cent of the population as indigenous, indigenous organisations state that the figure is nearer 60 per cent. Censuses on indigenous peoples are often unreliable and populations of indigenous peoples can be significantly different over ten-year periods if the formulation of the census question is changed. The latest estimate of the population of Guatemala is 15.5 million with an indigenous population of more than 6 million.

13 OEA/Ser.L/V/II.118 Doc, 5 rev. 1, 29 December 2003, Inter-American Commission on Human Rights' Thematic Report on Guatemala, *Justice and Social Inclusion: The Challenge of Democracy in Guatemala*, para. 213.

14 Ibid. See also 'Human Rights and Indigenous Issues: Mission to Guatemala', Report of the United Nations Special Rapporteur, Mr Rodolfo Stavenhagen, E/CN.4/2003/90/Add2, 10 February 2003, para. 5.

15 OEA/Ser.L/V/II.118 Doc, 5 rev. 1, 29 December 2003, para. 213, footnote 236.

16 Ibid, para. 214.

17 The 2008 Human Development Report estimates that 73 per cent of indigenous peoples are poor as opposed to 35 per cent for the Guatemalan population as a whole, and 26 per cent are extremely poor.

18 Hallman et al, *Multiple Disadvantages of Mayan Females: The Effects of Gender, Ethnicity, Poverty and Residence on Education in Guatemala*, Population Council Report (2006), p. 14.

19 Center for Economic and Social Rights (CESR), Fact Sheet No 3, p. 2, http://www2.ohchr.org/english/bodies/cedaw/docs/ngos/CESR_Guatemala43_en.pdf.

20 Report of James Anaya, Special Rapporteur on the situation of human rights and fundamental freedoms of indigenous peoples, 'Preliminary note on the application of the principle of consultation with indigenous peoples in Guatemala and the case of the Marlin mine', United Nations 2010 (A/HRC/15/37/Add.8), para. 5.

21 Observatory for the Protection of Human Rights, 'Guatemala "Smaller than David": the struggle of human rights defenders', International Fact-Finding Mission Report, February 2015, p. 13.

22 Guatemala ratified ILO Convention 169 in 1996, which is legally binding. Article 15 (2) reads: 'In cases in which the State retains the ownership of mineral or sub-surface resources or rights to other resources pertaining to lands, governments shall establish or maintain procedures through which they shall consult these peoples, with a view to ascertaining whether and to what degree their interests would be prejudiced, before undertaking or permitting any programmes for the exploration or exploitation of such resources pertaining to their lands. The peoples concerned shall wherever possible participate in the benefits of such activities, and shall receive fair compensation for any damages which they may sustain as a result of such activities.'

23 A/HRC/28/3/Add.1. Report of the United Nations High Commissioner for Human Rights on the activities of his office in Guatemala, 12 January 2015, para. 60. https://docs.un.org/en/A/HRC/28/3/Add.1

24 Although Mr Jangezoon told us that the application for the licence was for 115 metres, official sources from the Ministry informed us that it was for 120 metres.

25 Trans America Group owns other businesses as well, in the fields of telecommunications, technology, and financial and management consultancy; http://www.americatransgroup.com/.

26 OEA/Ser.L/V/II.118 Doc, 5 rev. 1, 29 December 2003, Inter-American Commission Thematic Report on Guatemala, *Justice and Social Inclusion: The Challenge of Democracy in Guatemala*, at para. 258. In its most recent report on Guatemala the United Nations High Commissioner for Human Rights refers that in protected natural areas, 'one hundred and seventy-four land conflicts [. . .] remain unsolved'. Report of the United Nations High Commissioner for Human Rights on the activities of his office in Guatemala, 12 January 2015, A/HRC.28/3/Add.1, para. 57.

27 Inter-American Commission Thematic Report on Guatemala, *Justice and Social Inclusion: The Challenge of Democracy in Guatemala*, ibid, at para. 258.

28 Report of the United Nations High Commissioner for Human Rights on the activities of her office in Guatemala, 13 January 2014, A/HRC/25/19/Add.1, at para. 59.

29 '*Que sí existen los Ch'ortí*'.

30 Manuel Scorza, *Garabombo el Invisible* (Editorial Planeta, 1972).

31 '*En cuanto a la legitimación de las personas jurídicas, la misma se ejerce a través de sus representantes legales, estas personas deben acreditar su existencia; su existencia de hecho o irregular no las legítima*', [Author's translation, emphasis added].

32 Article 20 of the Guatemalan Municipal code states: 'Article 20 Communities of indigenous peoples. The communities of indigenous peoples are forms of natural social cohesion and as such have the right to recognition of their juridical personality, needing to be entered in the registry office of the corresponding municipality, in respect of their internal organisation and administration which is governed according to their own norms, values and procedures, with their respective traditional authorities recognised and respected by the State, in accordance with constitutional and legal provisions' [Author's translation].

33 Registro de la Comunidad Indígena Maya Ch'orti' vecinos de la aldea de las Flores, del Municipio de Jocotán, departamento de

Chiquimula, Acta No 01-2014, Secretaría Municipal de Jocotán, Chiquimula, 5 August 2014.

34 Report of the United Nations High Commissioner for Human Rights on the activities of her office in Guatemala, 13 January 2014, A/HRC/25/19/Add.1, para. 59.

35 Report of the United Nations High Commissioner for Human Rights on the activities of her office in Guatemala, 7 January 2013, A/HRC/22/17/Add.1, para. 66.

36 Report of the United Nations High Commissioner for Human Rights on the activities of her office in Guatemala, 13 January 2014, A/HRC/25/19/Add.1, para. 59 citing Case file 266-2012, p. 18.

37 Report of the United Nations High Commissioner for Human Rights on the activities of his office in Guatemala, 12 January 2015, A/HRC.28/3/Add.1, para. 57.

38 Ibid.

39 For example, Article 57, section 7 of the Constitution of Ecuador guarantees 'free, prior and informed consultation, within a reasonable period of time, on plans and programmes for exploration, exploitation and sale of non-renewable resources located on their lands which could have environmental or cultural impacts on them.' It should be noted that although a number of states in the region recognize indigenous peoples' rights, they are less attentive in their implementation – a reality that has been characterised as the implementation gap.

40 In force since 1986 and amended in 1993. See Article 66.

41 Article 68.

42 Article 70.

43 Article 66.

44 Article 67: Protection of the Indigenous Agricultural Lands and Cooperatives.

The lands of the cooperatives, [the] indigenous communities or any other forms of communal or collective possession of agrarian ownership, as well as the family patrimony and the people's housing, will enjoy special protection of the State, [and] of preferential credit and technical assistance, which may guarantee their possession and development, in order to assure an improved quality of life to all of the inhabitants.

The indigenous communities and others that hold lands that historically belong to them and which they have traditionally administered in special form, will maintain that system [Author's translation and emphasis added].

45 Agreement on Identity and Rights of Indigenous Peoples, 1995, Article F 6 (c).

46 USAID Land Tenure And Property Profile on Guatemala, August 2010, available at http://usaidlandtenure.net/Guatemala.

47 Under Article 21 of the American Convention on Human Rights.

48 Inter-American Court of Human Rights, *Case of Yakye Axa Indigenous Community v Paraguay*, Judgment of 17 June 2005 (Merits, Reparations and Costs), para. 135.

49 Inter-American Court of Human Rights, *The Mayagna (Sumo) Awas Tingni Community v. Nicaragua*, para. 149.

50 The mission carefully revised all the material provided by the companies on the information dissemination of the projects and found no information as to risk assessments of possible negative aspects of such projects and how the projects planned to address or mitigate them.

51 A *piñata* is a decorated figure containing toys and sweets that is suspended from a height and broken open by blindfolded children as part of a celebration.

52 'Guiding Principles on Business and Human Rights: Implementing the United Nations 'Protect, Respect and Remedy' Framework', endorsed by the UN Human Rights Council in its resolution 17/4 of 16 June 2011. Prepared by the Special Representative of the Secretary-General on the issue of human rights and transnational corporations and other business enterprises. The Special Representative annexed the Guiding Principles to his final report to the Human Rights Council (A/HRC/17/31), which also includes an introduction to the Guiding Principles and an overview of the process that led to their development.

53 *'Es una tristeza cuando ellos entran.'*

54 *'Dicen que es para 'nos desarrolle.'*

55 *'Ellos los llaman desarrollo, cuando las máquinas pasan encima y ellos hablan de desarrollo.'*

56 *Nos dicen: 'sos un ruin y no sabes aprovechar el desarrollo.'*

57 *'No hay elegancia en la pobreza.'*
58 *'Desgano de superación, desgano por trabajar.'*
59 *'Tienen razón, en ningún idioma Maya la palabra 'desarrollo' existe. Sacar minerales, vender, tener papel moneda en banco no es nuestra forma de entender esta vida.'*
60 Victor Ferrigno F., 'El oscuro negocio de la luz', FLACSO, November 2009, http://issuu.com/flacsogt/docs/3epoca8.
61 OEA/Ser.L/V/II.118 Doc, 5 rev. 1, 29 December 2003, Inter-American Commission on Human Rights' Thematic Report on Guatemala, *Justice and Social Inclusion: The Challenge of Democracy in Guatemala*, para. 215.
62 See, for example, Comisión Internacional de juristas, *Empresas y violaciones a los derechos humanos en Guatemala: un desafío para la justicia*, December 2014.
63 The Marlin mine has been bedevilled with problems, social protests and criticisms by international bodies, including the UN and ILO. See, for example, the publication 'Metal mining and human rights: the Marlin Mine in San Marcos', Peace Brigades International, 2006.
64 For example, see http://www.mining.com/miners-in-guatemala-to-pay-ten-times-more-royalties-75059/.
65 The corridor is being sold as an alternative to the Panama Canal, which is unable to take the increased shipping of goods between the Pacific and Atlantic Oceans.
66 ILO Convention 169 requires governments to 'establish or maintain procedures through which they shall consult these peoples, with a view to ascertaining whether and to what degree their interests would be prejudiced, before undertaking or permitting any programmes for the exploration or exploitation of such resources pertaining to their lands', Article 15 (1).
67 UN Indigenous Declaration of 2007, Article 32 (2).
68 Constitutional Court of Guatemala, Case 3878-2007, Judgment of 21 December 2009, p. 24. The case concerned the '*El progreso*' cement company in San Juan Sacatepéquez (*San Juan Sacatepéquez* case). See: 'The consent and / or ratification of the provisions of multilateral documents listed above implies for the State of Guatemala, in short, the international commitment to

take a definite position on the right of consultation of indigenous peoples, expressed in several components: (i) its regulatory proper recognition and therefore the insertion into the constitutionality block or constitutional *corpus* as a fundamental right, by virtue of the provisions of Articles 44 and 46 of the Constitution; (ii) consequently, the obligation to ensure the effectiveness of the law in all cases where it is relevant; and (iii) the duty to make structural modifications required in the state apparatus – above all in terms of applicable legislation – to comply with this obligation according to the circumstances of the country.' [Author's translation].

69 '*Es grotesco lo que hacen acá.*'

70 '*Las Empresas son intocables.*'

71 '*Se intimida a la gente, o se usa el sistema penal de manera espuria con tipos penales propios del crimen tales como asociación ilícita, terrorismo, secuestro, sin considerar medidas sustitutiva.s*'

72 A/HRC/28/3/Add.1, op. cit., at para 46 [Author's emphasis].

73 '*A los abogados se les trata de terroristas.*'

74 OEA/Ser.L/V/II.118 Doc, 5 rev. 1, 29 December 2003, Inter-American Commission Thematic Report on Guatemala, *Justice and Social Inclusion: The Challenge of Democracy in Guatemala*, at para. 260.

75 Oscar García, 'Sala ordena restituir tierras a pueblo maya chortí de Jocotán', *Prensa Libre*, 17 July 2017.

4. All Mankind is One: The Camisea Gas Project in Peru and Non-contacted Tribal Peoples

1 Lewis Hanke, *All Mankind is One* (Northern Illinois University Press, 1974), p. xi.

2 Ibid.

3 Protected Area Watch, Nahua-Nanti Reserve, Perú, https://protectedareawatch.org/south-america/nahua-kupakagori-indigenous-reserve-peru/.

4 Ibid.

5 Survival International, Background Briefing, 'Camisea Gas Project', https://www.survivalinternational.org/about/Camisea.

6 Protected Area Watch, Nahua-Nanti Reserve, Perú, op. cit.

7 Bartolomé de las Casas, *A Short Account of the Destruction of the Indies* (Penguin, 1992), p. xxviii.

8 Inter-American Commission on Human Rights, 'Indigenous Peoples in Voluntary Isolation and Initial Contact in the Americas: Recommendations for the full respect of their Human Rights', OEA/Ser.L/V/II. Doc 47/13, 30 December 2013, para. 11. http://www.oas.org/en/iachr/indigenous/docs/pdf/report-indigenous-peoples-voluntary-isolation.pdf.
9 Ibid.
10 Ibid, at para. 1.
11 Ibid, at para. 2.
12 Ibid, para. 1.
13 Ibid.
14 Protected Area Watch, Nahua-Nanti Reserve, Perú, op. cit.
15 Forest Peoples Programme, *Vulnerando los derechos y amenazando vidas: el proyecto de gas de Camisea y los pueblos indígenas en aislamiento voluntario* (2014), p. 1.
16 Dan Collyns, 'Peru's indigenous people take battle over gas exploration to court', *Guardian*, 3 January 2013.
17 Jeremy Hance, 'Traveling the real and wild Dominican Republic', Mongabay, 29 January 2013.
18 Ibid.
19 Bartolomé de las Casas, *A Short Account of the Destruction of the Indies*, op. cit., p. xiii, Introduction.
20 Ibid, p. 3.
21 Ibid, p. 23.
22 Ibid, p. 24.
23 Ibid, p. 30.
24 Ibid, p. xiii, Introduction.
25 Ibid, p. 13.
26 Ibid.
27 Ibid, p. 7.
28 James Anaya, Relator Especial de las Naciones Unidas sobre los derechos de los pueblos indígenas, 'Observaciones sobre la ampliación de exploración y extracción de gas natural en el Lote 88 del proyecto Camisea', 24 March 2014, para. 2.
29 Supreme Decree No 28 028-2003-AG, https://assets.survivalinternational.org/documents/757/decreto-supremo-n-028-2003-ag.pdf.

30 Law No. 28736 for the Protection of Indigenous or Native Peoples in Isolation or Initial Contact.
31 Dan Collyns, 'Peru's indigenous people take battle over gas exploration to court', op. cit.
32 David Hill, 'UN urges Peru to Suspend US$480m Gas plans "immediately"', Huffpost, 19 March 2012.
33 Ibid.
34 'Piden evitar exterminio de indígenas aislados en reserve territorial', Servindi, 31 January 2014.
35 Forest Peoples Programme, 'Violating rights and threatening lives: The Camisea gas and indigenous peoples in voluntary isolation' (2014), p. 4.
36 Ibid.
37 Frederica Barclay and Pedro García Hierro, 'La Batalla por los Nanti', Informe 17, IWGIA, (2014), p. 7 [Author's translation].
38 Ibid.
39 Ibid.
40 See 1993 Peruvian Constitution, Chapter II (On Treaties), Article 55 reads: 'Treaties concluded by the State and in force, form part of national law' [Author's translation]. The State of Peru itself has informed the Inter-American Commission on Human Rights in its 'Response by the State of Peru to the Questionnaire for Consultation on indigenous Peoples in Voluntary Isolation and Initial Contact, received by the IACHR on June 4, 2013', that 'the treaties in force entered into by the State are part of domestic law.' See Inter-American Commission on Human Rights, 'Indigenous Peoples in Voluntary Isolation and Initial Contact in the Americas: Recommendations for the full respect of their Human Rights', OEA/Ser.L/V/II. Doc 47/13, 30 December 2013, at footnote 135.
41 See Peruvian Constitutional Tribunal, Judgment Case No. 0025-2005-PI/TC and 0026-2005-PI/TC, *Colegio de Abogados de Arequipa y otro*, 25 April 2006, paras 25–30.
42 See Peruvian Constitution, Final and transitory provisions, provision Fourth:

> 'The norms related to the rights and freedoms that the Constitution recognizes are interpreted in accordance with the

Universal Declaration of Human Rights and with the international treaties and agreements on the same matters ratified by Peru.' [Author's translation]

43 Peruvian Constitution, Article 89: 'The *Comunidades Campesinas* and native communities have a legal existence and are legal persons. They are autonomous in their organisation, in communal work and in the use and free disposal of their lands, as well as in economic and administrative matters, within the framework established by law. Ownership of their lands is imprescriptible, except in the case of abandonment provided for in the previous Article.' [Author's translation]. The term 'native communities' has been the ordinary way to refer to the indigenous peoples living in the Amazon.
44 Peruvian Constitution, Article 89: 'The State respects the cultural identity of the Peasant and Native Communities' [Author's translation].
45 Article 13.
46 Article 14.
47 Article 7.1.
48 Article 7.4
49 Article 15.
50 Ibid.
51 Under Article 15.2.
52 Inter-American Commission on Human Rights, 'Indigenous Peoples in Voluntary Isolation and Initial Contact in the Americas: Recommendations for the full respect of their Human Rights', op. cit., para. 6.
53 By virtue of Article 55 of the Peruvian Constitution. Peru ratified the ICCPR and the ICESCR on 28 April 1978.
54 Common Article 1.1 and 1.2 of the ICCPR and the ICESCR reads:

1. All peoples have the right of self-determination. By virtue of that right they freely determine their political status and freely pursue their economic, social and cultural development.
2. All peoples may, for their own ends, freely dispose of their natural wealth and resources without prejudice to any

> obligations arising out of international economic co-operation, based upon the principle of mutual benefit, and international law. In no case may a people be deprived of its own means of subsistence.

Reference to the right of self-determination is also made in the Preamble of ILO 169, relevant in understanding the very object and purpose of the said Convention.

55 Inter-American Commission on Human Rights, 'Indigenous Peoples in Voluntary Isolation and Initial Contact in the Americas: Recommendations for the full respect of their Human Rights', op. cit., para. 6, citing IACHR, Indigenous and Tribal Peoples' Rights over their Ancestral Lands and Natural Resources: Norms and jurisprudence of the Inter-American Human Rights System, OEA/Ser.L/V/II, 30 December 2009.

56 Ibid, at para. 21.

57 Inter-American Commission on Human Rights, 'Indigenous Peoples in Voluntary Isolation and Initial Contact in the Americas: Recommendations for the full respect of their Human Rights', ibid, para. 22

58 Oficina del Alto Comisionado de Derechos Humanos, UN, 'Directrices de Protección para los Pueblos Indígenas en Aislamiento y en contacto inicial de la Región Amazónica, el Gran Chaco y la Región Oriental del Paraguay', May 2012 [In Spanish only], at para. 66. This document is 'governed mainly by the principles of respect for the right of life and physical and cultural identity, the right to self-determination and no contact, and protection of the lands, territories, and natural resources traditionally occupied and used by indigenous peoples in voluntary isolation.', ibid, para. 53.

59 UN Permanent Forum on Indigenous Issues, Economic and Social Council, Report on the sixth session (14–25 May 2007), E/2007/43-E/C.19/2007/12, para. 39. Available at http://daccess-dds-ny.un.org/doc/UNDOC/GEN/N07/376/75/PDF/N0737675.pdf?OpenElement [Accessed 25 March 2015].

60 Inter-American Commission on Human Rights, 'Indigenous Peoples in Voluntary Isolation and Initial Contact in the Americas:

Recommendations for the full respect of their Human Rights', op. cit., footnote 42, citing the World Conservation Congress, Bangkok, Thailand, 17–25 November 2005.

61 Ibid, para. 22.

62 Inter-American Commission on Human Rights, 'Indigenous Peoples in Voluntary Isolation and Initial Contact in the Americas: Recommendations for the full respect of their Human Rights'; op. cit., para. 17, citing Ombudsperson Report NO 101, Office of the Ombudsperson of the Republic of Peru, Record No 2006-1282, Lima, January 2006, p. 51.

63 Ibid.

64 'The Last Man to Speak Taushiro in the Amazon', *New York Times*, https://www.youtube.com/watch?v=4qFG-sggtVc.

65 This was done within the framework of Decree Law No 22175 of 1978, Inter-American Commission on Human Rights, 'Indigenous Peoples in Voluntary Isolation and Initial Contact in the Americas: Recommendations for the full respect of their Human Rights', op. cit., para. 73. 8.

66 Ibid, para. 75.

67 UN Guidelines for the protection of indigenous peoples in voluntary isolation and initial contact of the Amazon region, Gran Chaco and Eastern Paraguay, § 52. Oficina del Alto Comisionado de Derechos Humanos, UN, 'Directrices de Protección para los Pueblos Indígenas en Aislamiento y en contacto inicial de la Región Amazónica, el Gran Chaco y la Región Oriental del Paraguay', May 2012 [Author's translation].

68 Law 28736, Article 1.

69 Law 28736, Article 5.

70 Article 5c [Author's translation].

71 [Author's translation].

72 Decree Law No 22175 of 1978.

73 Inter-American Commission on Human Rights, 'Indigenous Peoples in Voluntary Isolation and Initial Contact in the Americas: Recommendations for the full respect of their Human Rights'; op. cit., para. 74.

74 Supreme Decree No. 008-2007 MIMDES issued in 2007, which purported to regulate Law 28736, Decreto Supremo No

008-2007 MIMDES, 'Reglamento de la Ley para la Protección de Pueblos Indígenas u Originarios en Situación de Aislamiento y en Situación de Contacto Inicial', published in El Peruano, 5 October 2007.

75 Article modified by Novena Disposición Complementaria, Transitoria y Final del Decreto Supremo No 001-2012.

76 See Article 138 of the 1993 Peruvian Constitution.

77 'Gobierno de Alan García Pérez percibido como el 'más corrupto' según el CPI [Compañía Peruana de Estudios de Mercado]', *El Regional Piura*, 10 Diciembre 2017; Renato Cisneros, 'Alan Garcïa sí la debía, sí la temía', *The New York Times*, 17 April 2019.

78 Gideon Long, 'Garcia's death highlights Peru corruption scandals', *Financial Times*, 19 April 2019.

79 'Plata llega sola' le dijo Alan García, según Bayly, *AFP*, 9 December 2010.

80 Giovanna Castañedo Palomino, 'La Historia Inmobiliaria de Alan García', *El Comercio*, 21 October 2019.

81 Survival International, 'Peru: Mercury poisoning 'epidemic'sweeps tribe', 10 March 2016, https://www.survivalinternational.org/news/11167.

82 Forest Peoples' Programme, 'Violating rights and threatening lives: The Camisea Gas Project and Indigenous Peoples in Voluntary Isolation', January 2014, p. 6.

83 Roger Harrabin, 'Controversial gas from Peruvian Amazon arrives in UK', BBC, 4 March 2017.

84 Ibid.

85 Irene Banos Ruíz, 'Gas from the Amazon or fracked in Europe?', Deutsche Welle, 8 March 2017 https://www.dw.com/en/gas-from-the-amazon-or-fracked-in-europe/a-37834684.

86 'Organizations indígenas piden al Tribunal Constitucional declarar intangibilidad en área del Lote 88 de Camisea', *Agencia EFE*, 20 October 2021. The *Amparo* claim had been originally filed in 2013, https://gestion.pe/peru/organizaciones-indigenas-piden-al-tc-declarar-intangibilidad-area-del-lote-88-de-camisea-noticia/.

5. Sinking Islands: The Torres Strait Islanders

1 A position advanced by Philippe Sands, KC. See 'Climate Change and the Rule of Law: Adjudicating the Future in International Law', Public Lecture, 17 September 2015, p. 9.

2 BIICL and Institute of Small and Micro States Conference, 6–7 September 2018. See the presentation here: https://view.officeapps.live.com/op/view.aspx?src=https%3A%2F%2Fwww.twentyessex.com%2Fwp -content%2Fuploads%2F2024%2F02%2FClimate_Change_International_Dispute_Resolution.pptx&wdOrigin=BROWSELINK [accessed 16 February 2018]; for the recorded proceedings of the presentation, see https://www.youtube.com/watch?v=PJ4vzeuyKro&list=PL-M5JGQDtWUy9MBAvBm8_-D2tpgAZEbUw&index=4 [accessed 16 February 2018].

3 There had been a previous attempt at taking the climate as a human rights issue before the Inter-American Commission on Human Rights, which had unfortunately failed. Sometime after my presentation in the Conference, I found slides from an illuminating lecture on the law of the sea in Singapore, in which Professor Alan Boyle also considered that climate degradation fell within the Convention on the Law of the Sea remit.

4 Adopted by 196 parties at the UN Climate Change Conference (COP21) in Paris, France, on 12 December 2015. It entered into force on 4 November 2016.

5 Monica Feria-Tinta, 'The Role of International Law and Arbitration in Enforcing the Paris Agreement', Kluwer Arbitration Blog, 31 December 2016, https://arbitrationblog.kluwerarbitration.com/2016/12/31/the-role-of-international-law-and-arbitration-in-enforcing-the-paris-agreement/ [accessed 22 December 2023].

6 '13 Islands that will disappear in the next 80 years', *Readers Digest*, 14 June 2022.

7 See Monica Feria-Tinta's criticisms of such a position in a 2018 presentation at a conference organised by BIICL and Institute of Small and Micro States Conference, 6–7 September 2018. For the recorded proceedings of the presentation, see https://www.youtube.com/watch?v=PJ4vzeuyKro&list=PL-M5JGQDtWUy9MBAvBm8_-D2tpgAZEbUw&index=4 [accessed 16 February 2018].

8 See Keith Pabai's statement, as I collated it, and cited it

in Communication under the Optional Protocol to the International Covenant on Civil and Political Rights, 13 May 2019 ('Communication to the UN Human Rights Committee'), p. 7, https://ourislandsourhome.com.au/wp-content/uploads/sites/92/2021/03/Billy-v-Australia_Communication-under-the-Optional-Protocol-to-the-ICCPR_13.5.19_Redacted.pdf.

9 IMO, Resolution MEPC. 133 (53), adopted on 22 July 2005, Annex 21, para. 2.1.

10 Ibid, Annex 21, para. 2.1.

11 Ibid.

12 Australian Bureau of Statistics, '2016 Census shows growing Aboriginal and Torres Strait Islander population', available at: http://www.abs.gov.au/ausstats/abs@.nsf/MediaRealesesByCatalogue/02D50FAA9987D6B7CA25814800087E03 [accessed 1 March 2019].

13 There are four main island groups, with increased linguistic and cultural similarities within each group, namely: North-Western (Boigu, Saibai, Dauan), Western (Badu, Moa, Mabuiag), Central (Masig, Poruma, Warraber, Iama) and Eastern (Erub, Mer, Ugar).

14 Academic sources of information about the culture include Anna Shnukl, 'The post-contact created environment in the Torres Strait Central Islands' Memoirs of the Queensland Museum, *Cultural Heritage Series* (2004), 3(1), 317–346, Brisbane, available at https://www.museum.qld.gov.au/collections-and-research/memoirs/culture-3/mqm-c3-1-13-shnukl, and J. Beckett, *Torres Strait Islanders: Custom and Colonialism* (Cambridge University Press, 1987).

15 John Cordell and Judith Fitzpatrick, *'Torres Strait: Cultural Identity and the Sea'*. *Cultural Survival*, 19 February 2010, https://www.culturalsurvival.org/publications/cultural-survival-quarterly/torres-strait-cultural-identity-and-sea

16 Ibid.

17 Ântonio Cançado Trindade's words in 'Conversación con Antonio Augusto Cançado Trindade. Reflexiones sobre la Justicia Internacional', Valencia, 2013, cited by SE Philippe Couvreur, 'In Memoriam: Ântonio Cançado Trindade (1947–2022)', *Diplomat Magazine*, 23 September 2023 [Author's translation].

18 See Keith Pabai's statement, as I collated it, and cited it in Communication to the UN Human Rights Committee, op. cit., p. 12.

19 Ibid.

20 Ted Billy's statement as I read it when looking into the evidence in this case. [Author's emphasis].

21 UN Environment Programme, 'The German Islet putting wind into the sails of climate action and clean seas', 29 October 2019, https://www.unep.org/news-and-stories/story/german-islet-putting-wind-sails-climate-action-and-clean-seas#:~:text=Sylt%20is%20vulnerable%20to%20sea,The%20island%20is%20shrinking.

22 Ted Billy's statement (at para. 25), as I collated it, and cited it in Communication to the UN Human Rights Committee, op. cit., p. 11.

23 Yessie Mosby's statement, as I collated it and cited it in Communication to the UN Human Rights Committee, op. cit., p. 11.

24 Ibid.

25 Ibid.

26 As I read in Nazareth Warria's statement.

27 IMO, Resolution MEPC. 133 (53), adopted on 22 July 2005, para. 2.3.

28 Ibid.

29 Keith Pabai's statement, as I collated it and cited it in Communication to the UN Human Rights Committee, op. cit., p. 12.

30 Nazareth Warria's statement, as I read it when looking into the evidence in this case at para. 20.

31 Nazareth Warria's statement, as I collated it and cited it in Communication to the UN Human Rights Committee, op. cit., p. 12.

32 Kabay Tamu's statement, as I collated it and cited it in Communication to the UN Human Rights Committee, op. cit., p. 12.

33 Ibid.

34 *Sheila Watt-Cloutier et al v United States*, Inter-American Commission on Human Rights, Petition Seeking Relief from Violations resulting from Global Warming caused by Acts and Omissions of the United States (7 December 2005). See 'Petition Seeking Relief from Violations resulting from Global Warming caused by Acts and Omissions of the United States' (Grantham Research Institute on Climate Change and the Environment), https://climate-laws.org/cclow/geographies/9/litigation_cases/7052.

35 Acceptance speech, Sheila Watt-Cloutier, Right livelihood award.
36 *Sheila Watt-Cloutier et al v United States*, Application, p. 1.
37 Ibid, p. 37.
38 Ibid.
39 Ibid.
40 Ibid, p. 61.
41 Yessie Mosby's statement, as I collated it and cited it in Communication to the UN Human Rights Committee, op. cit., p. 13, referring to §§ 90–91 of his statement.
42 J. Johnson and D. Welch, 'Climate change implications for Torres Strait fisheries: assessing vulnerability to inform adaptation', *Climatic Change* (2016), 135(3–4), 611–624 DOI 10.1007/s10584-015-1583-z.
43 Stanley Marama's statement, as I collated it and cited in Communication to the UN Human Rights Committee, op. cit., p. 6, referring to §§ 34–35 of his statement.
44 Stanley Marama's statement in the case as I read it when looking into the evidence in this case.
45 Ted Billy's statement, as I collated it and cited it in Communication to the UN Human Rights Committee, op. cit., p. 6, referring to § 31 of his statement.
46 Ibid, §§ 26–27.
47 See https://www.pzja.gov.au/resources/dugong-and-turtle-fisheries.
48 Australian Government, Department of Climate Change, Energy, the Environment and Water, https://www.dcceew.gov.au/environment/marine/marine-species/marine-turtles#:~:text=Under%20the%20Native%20Title%20Act,native%20title%20rights%20and%20interests.
49 Leah Lui-Chivizhe, 'Torres Strait Islanders Women, turtle hunting and spirituality', 13 September 2022, https://www.broadagenda.com.au/2022/torres-strait-islander-women-turtle-hunting-and-spirituality/.
50 Ibid.
51 Ibid.
52 Ibid.
53 As I read in Ted Billy's statement.

54 Kabay Tamu's statement, as I collated it and cited it in Communication to the UN Human Rights Committee, statement, op. cit., p. 6, referring to § 26 of his statement.

55 TSRA, 'Torres Strait Climate Change Strategy 2014–18: Building Community Adaptive Capacity and Resilience', p. iii, https://www.tsra.gov.au/wp-content/tsra-archive/data/assets/pdf_file/0003/6393/TSRA-Climate-Change-Strategy-2014-2018-Upload.pdf .

56 Katharine Murphy, 'Scott Morrison brings coal to question time: what fresh idiocy is this?', *Guardian*, 9 February 2017.

57 Ibid.

58 Philippe Sands KC in 'Climate Change and the Rule of Law', Public Lecture, 17 September 2015, op. cit., p. 9.

59 Ibid.

60 Rosalyn Higgins, *Problems and Process: International Law and How we use it* (OUP, 1994), p. 9.

61 See Torres Strait Islanders case, Communication under the Optional Protocol to the International Covenant on Civil and Political Rights, 13 May 2019, paras 131–135, available at https://climatecasechart.com/wp-content/uploads/non-us-case-documents/2019/20190513_CCPRC135D36242019_complaint.pdf? [Accessed 1 January 2024].

62 NITV News, 'Funding to build seawalls in the Torres Strait, amidst calls for climate action', 22 December 2019.

63 For an analysis of the decision, see Monica Feria-Tinta, 'Torres Strait Islanders: United Nations Human Rights Committee Delivers Ground-Breaking Decision on Climate Change Impacts on Human Rights', EJIL Talk!, 27 September 2022.

64 Ibid.

6. West Timor: The Montara Matara Oil Spill Case

1 While conventions like the International Convention on Oil Pollution Preparadness, Response and Cooperation (OPRC), dealing with measures to prevent pollution from ships and the 1969 International Convention on Civil Liability for Oil Pollution Damage (adopted under the auspices of the International Maritime Organisation) seeking to ensure that adequate compensation was

paid to victims and the liability was placed on the shipowner, these conventions all deal with pollution relating to the shipping industry. They do not cover platforms. In 2003, an additional Protocol on the Establishment of a Supplementary Fund for Oil Pollution Damage established an International Oil Pollution Compensation Supplementary Fund, in order to provide an additional third tier compensation for oil pollution damage (also in relation to the shipping industry) was adopted.

2 See, for example, K. S. Abraham, 'Catastrophic Oil Spills and the Problem of Insurance', *Vanderbilt Law Review*, 64(6) 1769–1791 (2011).

3 Ibid, p. 1771.

4 J. Kollewe and T. Macalister, 'Arctic oil rush will ruin ecosystem, warns Lloyd's of London', *Guardian*, 12 April 2012.

5 Ibid.

6 Nina Lakhani, '"We've been abandoned": a decade later, Deepwater Horizon still haunts Mexico', *Guardian*, 19 April 2020.

7 Ibid.

8 N. Janowitz, 'Revealed: Documents show BP quietly paid just $25 million to Mexico after the Worst Oil Spill of the Century', Buzzfeed.News, 28 September 2018.

9 C. R. Payne, 'Developments in the Law of Environmental Reparations: A case study of the UN Compensation Commission', in C. Stahn, J. Iverson & J. Easterday, (eds), *Environmental Protection and Transitions from Conflict to Peace: Clarifying Norms, Principles, and Practices*, (OUP, 2017) pp. 329–366.

10 Report of the Special Rapporteur on the issue of human rights obligations relating to the enjoyment of a safe, clean, healthy and sustainable environment, 'Human Rights Obligations relating to the enjoyment of a safe, clean, healthy, and sustainable environment', A/74/161, 15 July 2019, §43.

11 Report of the Special Rapporteur on the implications for human rights of the environmentally sound management and disposal of hazardous substances and wastes, A/74/480, 7 October 2019, §7.

12 Ibid.

13 The regencies on whose behalf I made the communication were: City of Kupang; Kupang Regency; Rote Ndao Regency; Timor

Tengah Selatan Regency; Timor Tengah Utara Regency; Belu Regency; Alor Regency; Lembata Regency; East Flores Regency; East Sumba Regency; Central Sumba Regency; West Sumba Regency; and Sabu Raijua Regency.

14 Report by the Montara Commission of Inquiry, led by Commissioner David Borthwick, 17 June 2010, p. 36.
15 Report by the Montara Commission of Inquiry, p. 5.
16 Ibid.
17 Report by the Montara Commission of Inquiry, led by Commissioner David Borthwick, 17 June 2010.
18 Ibid, p. 5.
19 Ibid, p. 26.
20 'After the Spill: Investigating Australia's Montara oil disaster in Indonesia', Australian Lawyers Alliance, July 2015, p. 225.
21 Report by the Montara Commission of Inquiry, p. 6 [Author's emphasis].
22 Ibid, p. 11.
23 Ibid, p. 11.
24 Ibid, p. 5.
25 Ibid, p. 11.
26 Bills and Agostini, *Offshore Petroleum Safety Regulation*, p. xi, cited in Report by the Montara Commission of Inquiry, p. 161, footnote 245. https://www.industry.gov.au/sites/default/files/2022-09/montara-commission-of-inquiry-report-june-2010.pdf.
27 Report by the Montara Commission of Inquiry, p. 6.
28 Ibid, p. 11.
29 Ibid, p. 14.
30 'After the Spill: Investigating Australia's Montara oil disaster in Indonesia', op. cit., p. 4.
31 Montara Oil Spill Inquiry Analysis – Oil Spill Response, Report to WWF Australia, prepared by Nuka Research & Planning Group LLC, February 2010, p. 4.
32 Ibid, p. 18.
33 Edward O. Wilson, *Half-Earth* (Liveright, 2016), p. 68.
34 Ibid, p. 69.
35 Montara Oil Spill Inquiry Analysis – Oil Spill Response, Report to WWF Australia, op. cit., p. 19.

36 'After the Spill: Investigating Australia's Montara oil disaster in Indonesia', p. 59.
37 Ibid.
38 Ibid, p. 4.
39 Ibid, p. 4.
40 Ibid, see in particular pp. 45–105.
41 Ibid, p. 5.
42 Ibid, p. 7.
43 Ibid, p. 227.
44 Rachel Siewert, Senator for Western Australia, letter addressed to The Honorable Maritime Affairs Coordinating Minister Public Indonesia, Mr Luhut Binsar Pandjaitan. [undated, contemporary to the 8th Anniversary of the Montara tragedy]
45 After the Spill: Investigating Australia's Montara oil disaster in Indonesia', op. cit., p. 228.
46 Letter of Ward Keller, Australian lawyer acting for the West Timor communities, addressed to The Honourable Malcolm Turnbull, Prime Minister of Australia, 7 January 2016.
47 Inter-American Commission on Human Rights, Advisory Opinion, 23/17, 15 November 2017, § 47.
48 Inter-American Commission on Human Rights (IAComHR), *Indigenous and Tribal People's Rights over Their Ancestral Lands and Natural Resources, Norms and Jurisprudence of the Inter-American Human Rights System*, OEA/Ser.I/L/V/II. Doc 56/09 (30 December 2009), § 192.
49 Inter-American Commission on Human Rights, Advisory Opinion, 23/17, 15 November 2017, § 47.
50 See ibid, § 124.
51 The Special Rapporteur on the implications for human rights of the environmentally sound management and disposal of hazardous substances and wastes.
52 The Special Rapporteur on the issue of human rights obligations relating to the enjoyment of a safe, clean, healthy and sustainable environment; the Special Rapporteur on the right to food; the Special Rapporteur on the rights of indigenous peoples; and the Special Rapporteur on extreme poverty and human rights.

53 Amanda Battersby, 'PTTEP to pay $129 million compensation for 2009 Montara oil spill', *Upstream*, 22 November 2022, https://www.upstreamonline.com/safety/pttep-to-pay-129-million-compensation-for-2009-montara-oil-spill/2-1-1358744.

7. The Heart of the Earth: La Línea Negra

1 T. Barnett and H. Vilchez, 'The Arhuacos: A message from the Mamos, the Prophets of the Sierra Nevada', Pulitzer Centre, 2 November 2021.
2 Ibid.
3 Ibid.
4 Smithsonian, 'Earth's highest coastal mountain on the move', 20 September 2010.
5 Smithsonian Insider, 'Earth's highest coastal mountain range moved 1,367 miles in 170 million years', *Research News Science and Nature*, 23 September 2010.
6 Radio Nacional en Colombia, 'Gobierno Colombiano reconoce "la línea negra" como espacio sagrado indígena', 15 August 2023.
7 Consejo Territorial de Cabildos Indígenas de la Sierra Nevada de Santa Marta, 'Hacia una Política Pública Ambiental del Territorio Ancestral de la Línea Negra de los Pueblos Iku, kággaba, Wiwa and Kankuamo de la Sierra Nevada de Santa Marta en la Construcción Conjunta con Parques Nacionales Naturales', May 2020.
8 *Aluna: A Journey to Save the World*, documentary by Alan Ereira, 2011. https://www.youtube.com/watch?v=ftFbCwJfs1I.
9 Ibid.
10 Ibid.
11 Ibid.
12 Ibid.
13 Global Conservation, GC Mission to Sierra Nevada and Tayrona National Parks, Colombia, 25 February 2023, https://globalconservation.org/news/gc-mission-sierra-nevada-and-tayrona-national-parks-colombia/.
14 IUCN, 'Scientists identify the world's most irreplaceable protected areas', 14 November 2013, https://www.iucn.org/content/scientists-identify-worlds-most-irreplaceable-protected-areas.

15 Selva, 'Aves endémicas de la Sierra Nevada de Santa Marta', https://www.selva.org.co/programas/ciencia-de-la-conservacion/aves-endemicas-de-la-sierra-nevada-de-santa-marta/ [accessed 4 July 2020; Author's translation].
16 Consejo Territorial de Cabildos Indígenas de la Sierra Nevada de Santa Marta, Documento Madre de la Línea Negra – Jaba Séshizha – de los Cuatro Pueblos Indígenas de la Sierra Nevada de Santa Marta'. 9 December 2015, https://justiciaambientalcolombia.org/wp-content/uploads/2022/08/DOCUMENTO-MADRE-Linea-Negra-Final-Dic10_2015.pdf.
17 Ibid.
18 Escrito de los pueblos Wiwa and Kankuamo de la Sierra Nevada de Santa Marta a la Corte Interamericana de Derechos Humanos – Opinión Consultiva sobre Emergencia Climática y Derechos Humanos presentadas por los Estados de Colombia y Chile https://corteidh.or.cr/sitios/observaciones/OC-32/5_kankuamo_wiwa.pdf.
19 A Ramsar site is a wetland site designated to be of international importance under the Ramsar Convention, or 'The Convention on Wetlands', an international environmental treaty signed on 2 February 1971 in Ramsar, Iran, under the auspices of UNESCO.
20 Escrito de los pueblos Wiwa and Kankuamo de la Sierra Nevada de Santa Marta a la Corte Interamericana de Derechos Humanos – Opinión Consultiva sobre Emergencia Climática y Derechos Humanos presentadas por los Estados de Colombia y Chile, op. cit.
21 Consejo Territorial de Cabildos Indígenas de la Sierra Nevada de Santa Marta, Documento Madre de la Línea Negra – Jaba Séshizha – de los Cuatro Pueblos Indígenas de la Sierra Nevada de Santa Marta', 9 December 2015, op. cit.
22 Namely Judgment T-547 of 2010.
23 In Order 004/09.
24 Mr Arturo Blanco Ordóñez, a Colombian businessman, was the founder of Puerto Brisa and the Prodeco company. Prodeco is today a wholly owned Colombian subsidiary of Glencore plc engaged in the exploration, production, transportation and shipping of thermal and coking coal, and related infrastructure.
25 Principle 22, Rio Declaration on Environment and Development, https://www.cbd.int/doc/ref/rio-declaration.shtml.

26 Principle 23, Ibid.
27 Article 5(1) of the Paris Agreement.
28 *Amicus Curiae*, of ABColombia and Colombian Caravana, En la Demanda de Nulidad Simple del Decreto 1500 Línea Negra (Sierra Nevada de Santa Marta), 30 June 2020.
29 'The Arhuaco peoples bless Colombian president-elect Petro', *Telesur*, 5 August 2022.
30 Escrito de los pueblos Wiwa and Kankuamo de la Sierra Nevada de Santa Marta a la Corte Interamericana de Derechos Humanos – Opinión Consultiva sobre Emergencia Climática y Derechos Humanos presentadas por los Estados de Colombia y Chile, op. cit.
31 Christopher P. Baker, 'The ancient guardians of the Earth', BBC, 3 April 2019.

8. The Masewal and the Constitutionality of Federal Mining Law in Mexico

1 CEMDA, *Entre la Tierrita y el Suelo*, p. 29.
2 Alessandro Questa, 'Mountains in resistance. The Masewal world-view in the face of climate change and extractivism', Cuicuilco. *Rev. cienc. antropol*, 25(72), Ciudad de México May/August 2018 [Author's translation].
3 Códice Masewal, Plan de Vida, 2027, p. 24, https://patrimoniobiocultural.com/subidas/2022/06/PARTE-1-CÓDICE-MASEWAL-2022.pdf [Author's translation].
4 Ibid, p. 31.
5 CEMDA, *Entre la Tierrita y el Suelo*, p. 29.
6 Ibid, p. 32.
7 Alessandro Questa, 'Mountains in resistance. The Masewal world-view in the face of climate change and extractivism', op. cit. p. 128 [Author's translation]
8 Ibid, p. 127 [Author's translation].
9 Ibid [Author's translation].
10 Ibid [Author's translation].
11 Ibid [Author's translation].
12 Yuribia Velásquez Galindo & Hugo Rodríguez Gonzalez, 'El agua y sus significados: una aproximación al mundo de los Nahuas en

México', *Antípoda*. Revista de Antropología y Arqueología 34, (2019) 69–88, p. 80. [Author's translation]

13 Ibid.

14 Ibid.

15 Edward O. Wilson, *Half-Earth* (Liveright, 2016), p. 77.

16 Ibid, p. 164.

17 BBC, *War on Plastic with Hugh and Anita*, 5 June 2019, https://www.bbc.co.uk/programmes/p07c90ff.

18 EcoWatch, 'Chile's Atacama Desert: Where Fast Fashion Goes to Die', 15 November 2021.

19 Ibid.

20 Alessandro Questa, 'Broken Pillars of the Sky: Masewal Actions and Reflections on Modernity, Spirits, and a Damaged World', in Rosalyn Bold (ed.), *Indigenous Perceptions of the End of the World, Creating a Cosmopolitics of Change* (Palgrave Macmillan, 2019), p. 44.

21 Alessandro Questa, 'Mountains in resistance. The Masewal world-view in the face of climate change and extractivism', op. cit.

22 Ibid. [Author's translation]

23 Ibid. [Author's translation]

24 Bill Porter, *Road to Heaven, Encounters with Chinese Hermits* (Counterpoint, 1993), p. 23.

25 Alessandro Questa, 'Mountains in resistance. The Masewal world-view in the face of climate change and extractivism', op. cit. [Author's translation].

26 Ibid.

27 David Hinton (ed. and translator), *Mountain Home: The Wilderness Poetry of Ancient China* (Anvil Press Poetry, 2007).

28 Ibid, p. xiii.

29 From *Endless River: Li Po and Tu Fu, A Friendship in Poetry*, translated by Sam Hamill (Weatherhill, 1993).

30 Aitareya Upanishads, *The Upanishads*, translation by Swammi Nikhilananda, vol. 3 (Ramakrishna-Vivekananda Center, N.Y. 2003) (fourth edition), p. 21.

31 Ibid, p. 24.

32 Cormac Cullinan, *Wild Law*, op. cit.

33 Ibid, pp. 82–83.

34 Comunicado de Prensa No. 009/2021, 'La Segunda Sala de la SCJN ratifica el derecho de consulta previa de los pueblos indígenas, en específico, del pueblo Maseual', https://www.internet2.scjn.gob.mx/red2/comunicados/noticia.asp?id=6315.
35 CEMDA, 'Pueblo Masewal gana lucha en contra de la minería', 18 March 2022.
36 Ibid.
37 Alessandro Questa, 'Broken Pillars of the Sky: Masewal Actions and Reflections on Modernity, Spirits, and a Damaged World', op. cit., p. 33.
38 Ibid, p. 46.
39 Ibid. p. 40.
40 Ibid. p. 41.
41 Ibid.
42 Ibid, p. 46.
43 Alessandro Questa Rebolledo, 'Dancing with the Spirits', *Anthropology News*, 16 April 2019.
44 Kitxpala, 'Danza de Wewentiyo', http://www.sanildefonso.org.mx/expos/kixpatla/territorios-37-masewal-nahua.html.
45 Alessandro Questa Rebolledo, 'Dancing Spirits: Towards a Masewal ecology of interdependence in the northern highlands of Puebla', PhD thesis, Anthropology Department, University of West Virginia 18 November 2017, https://libraetd.lib.virginia.edu/public_view/q811kj959.
46 Ibid.
47 Gamaliel Churata, *El Pez de Oro* (Cátedra, Letras Hispánicas, 2012).
48 Gamaliel Churata, 'Notas para la historia del pensamiento boliviano', in Cuadernos Literarios I (28), Suplemento de *Última Hora, 4/8/1949*, cited in Mauro Mamani Macedo, 'Ahayu-Watan: una categoría andina para explicar nuestra cultural', Caracol 9/Dossié 93–127, p. 97 [Author's translation].
49 Ibid. [Author's translation]

9. The Rights of Nature Case: Los Cedros Cloud Forest

1 Judgment of 19 June 2019, *Amicus Curiae* of Fred Larreátegui Fabara, p. 17.

2 Article 10 of the Constitution of Ecuador recognises Nature as the holder of constitutional rights. It reads: 'Nature will be subject to those rights recognized by the Constitution.'
3 Article 71 [Author's translation].
4 Julian Hoffman, *Irreplaceable*, (Penguin, 2020), p. 11.
5 Ibid, p. 12.
6 Ibid.
7 Roy et al, 'New Mining Concessions Could Severely Decrease Biodiversity and Ecosystem Services in Ecuador', *Tropical Conservation Science*, (2018), 11, 1–20, p. 1.
8 Ibid, p. 3, referring to N. Myers, R. A. Mittermeier, C. G. Mittermeier, G. A. B. da Fonseca and J. Kent, 'Biodiversity hotspots for conservation priorities', *Nature*, (2000), 403, 853–858.
9 Roy et al, 'New Mining Concessions Could Severely Decrease Biodiversity and Ecosystem Services in Ecuador', op. cit., relying on J. H. Brown, 'Why are there so many species in the tropics', *Journal of Biogeography*, (2014), 41, 8–22.
10 W. I. Eiserhardt, T. I. P. Couvreur and W. J. Baker, 'Plant phylogeny as a window on the evolution of hyperdiversity in the tropical rainforest biome', *New Phytologist*, (2017), 214(4), 1408–1422.
11 Roy et al, 'New Mining Concessions Could Severely Decrease Biodiversity and Ecosystem Services in Ecuador', op. cit., p. 6, relying on P. M. Jørgensen and S. León-Yáñez, *Catalogue of the Vascular Plants of Ecuador* (Missouri Botanical Garden Press, 1999).
12 See https://es.wikipedia.org/wiki/Bosque_nuboso.
13 Roy et al, 'New Mining Concessions Could Severely Decrease Biodiversity and Ecosystem Services in Ecuador', op. cit., p. 2.
14 Ibid, relying on Eiserhardt et al, 2017; C. Hughes and R. Eastwood, 'Island radiation on a continental scale: Exceptional rates of plant diversification after uplift of the Andes', *Proceedings of the National Academy of Sciences of the United States of America, 103(27): 10334-10339*, 2006.
15 Roy et al, 'New Mining Concessions Could Severely Decrease Biodiversity and Ecosystem Services in Ecuador', op. cit., p. 3, relying on Jørgensen and León-Yáñez, 1999; S. Leon-Yáñez, R. L. Valencia, N. Pitman, L. Endara, C. Ulloa-Ulloa and H. Navarrete, *Libro Rojo de las Plantas Endémicas del Ecuador* (Pontificia

Universidad Católica del Ecuador, second edition, 2012); R. Valencia, N. Pitman, S. León-Yáñez and P. M. Jørgensen (eds), *The Red Book of the Endemic Plants of Ecuador 2000* (Publications of QCA, Herbarium, Pontificia Universidad Católica del Ecuador, 2000).

16 Roy et al, 'New Mining Concessions Could Severely Decrease Biodiversity and Ecosystem Services in Ecuador', op. cit., p. 4, relying on F. Borchsenious, 'Patterns of plant species endemism in Ecuador', *Biodiversity and Conservation*, (1997), 6(3), 379–399; N. C. A. Pitman and P. M. M. Jørgensen, 'Estimating the size of the world's threatened flora', *Science*, (2002), 298(5595): 989. doi:10.1126/science.298.5595.989

17 Roy et al, 'New Mining Concessions Could Severely Decrease Biodiversity and Ecosystem Services in Ecuador', op. cit., p. 1.

18 Roy et al, 'New Mining Concessions Could Severely Decrease Biodiversity and Ecosystem Services in Ecuador', op. cit., p. 3 relying on H. Kreft and W. Jetz, 'Global patterns and determinants of vascular plant diversity', *Proceedings of the National Academy of Sciences of the United States of America*, (2007), 104(4), 5925–5930.

19 W. I. Eiserhardt, T. I. P. Couvreur and W. J. Baker, 'Plant phylogeny as a window on the evolution of hyperdiversity in the tropical rainforest biome', *New Phytologist*, (2017), 214(4), 1408–1422.

20 Roy et al, p. 3, relying on H. Kreft and W. Jetz, 'Global patterns and determinants of vascular plant diversity', *Proceedings of the National Academy of Sciences of the United States of America*, (2007), 104(14), 5925–5930.

21 Edward O. Wilson, *Half-Earth* (Liveright, 2016), p. 50.

22 Edward O. Wilson, *Biophilia* (Harvard University Press, 1984), p. 8.

23 Judgment of 19 June 2019, summing up the *Amicus Curiae* of the biologist Elisa Levy Ortiz, pp. 66–67.

24 Roy et al, 'New Mining Concessions Could Severely Decrease Biodiversity and Ecosystem Services in Ecuador', op. cit., p. 7.

25 Ibid.

26 Ibid, relying on C. R. Hutter and J. M. Guayasamin, 'Cryptic diversity concealed in the Andean cloud forests: Two new species of rainfrogs (*Pristimantis*) uncovered by molecular and bioacoustics data', *Neotropical Biodiversity*, (2015), 1(1), 36–59.

27 Roy et al, 'New Mining Concessions Could Severely Decrease Biodiversity and Ecosystem Services in Ecuador', op. cit., p. 7.
28 Ibid, p. 8.
29 Judgment, 19 June 2019, summing up the *Amicus Curiae* of biologist Elisa Levy Ortiz, p. 66.
30 Roy et al, 'New Mining Concessions Could Severely Decrease Biodiversity and Ecosystem Services in Ecuador', op. cit., p. 14.
31 Roy et al, 'New Mining Concessions Could Severely Decrease Biodiversity and Ecosystem Services in Ecuador', op. cit., p. 15, relying on E. Deharo, M. Sauvain, C. Moretti, B Richard, E. Ruiz and G. Massiot, 'Antimalarial effect of n-hentriacontanol isolated from Cuatresia sp. (Solanaceae)', *Annales de Parasitologie Humaine et Comparée*, (1992), 67(4), 126–127; M. Krugliak, E. Deharo and G. Shalmiev 'Antimalarial effects of C18 Fatty-acids on Plasmodium falciparum in culture and on Plasmodium vinckei petteri and Plasmodium yoelii nigeriensis in vivo', *Experimental Parasitology*, (1995), 81(1), 97–105.
32 Acción de Protección Sala Multicompetente de la Corte Provincial de Imbabura (Demandado, Procuradoría General del Estado y Ministerio del Ambiente), Judgment of 19 June 2019, p. 3 [Author's translation].
33 Ibid.
34 Ibid.
35 Ibid, p. 4.
36 Ibid. p. 3.
37 'Oso Andino, Especie en peligro de extinción en el Ecuador', https://www.arcgis.com/apps/MapJournal/index.html?appid=66952e6b252f4cfda1037f9cdf0acf2b.
38 Ibid.
39 Ibid.
40 Ibid.
41 Wilson, *Half-Earth*, op. cit., p. 54.
42 Wilson, *Biophilia*, op. cit., p. 24.
43 Ibid.
44 Ibid.
45 Roy et al, 'New Mining Concessions Could Severely Decrease Biodiversity and Ecosystem Services in Ecuador', op. cit., p. 5.

46 Ibid, p. 1.
47 Ibid, p. 13.
48 Ibid, relying on N. M. Haddad, L. A. Brudvig, J. Clobert, K. F. Davies, A. Gonzalez, R. D. Holt and J. R. Townshend, 'Habitat fragmentation and its lasting impact on Earth's ecosystems', *Science Advances*, (2015), 1(2). doi:10.1126/sciadv.1500052
49 Roy et al, 'New Mining Concessions Could Severely Decrease Biodiversity and Ecosystem Services in Ecuador', op. cit., p. 13.
50 Ibid, relying on S. Báez, L. Jaramillo, F. Cuesta and D. A. Donoso, 'Effects of climate change on Andean biodiversity: A synthesis of studies published until 2015', *Neotropical Biodiversity*, (2015), 2(1), 181–194.
51 Roy et al, 'New Mining Concessions Could Severely Decrease Biodiversity and Ecosystem Services in Ecuador', op. cit., p. 13.
52 Ibid, relying on K. M. Jack, & F. A. Campos, 'Distribution, abundance, and spatial ecology of the critically endangered Ecuadorian capu chin (Cebus albifrons aequatorialis)'. *Tropical Conservation Science*, (2012), 5(2): 173–191; M. Peck, J Thorn; A. Mariscal, A. Baird, D. Tirira & D. Kniveton, 'Focusing conservation efforts for the critically endangered brown-headed spider monkey (Ateles fusciceps) using remote sensing, modeling, and playback survey methods. International Journal of Primatology', (2010), 32(1): 134–148.
53 Roy et al, 'New Mining Concessions Could Severely Decrease Biodiversity and Ecosystem Services in Ecuador', op. cit., p. 13, relying on J. A. de la Torre, J. González-Maya, H. Zarza, G. Ceballos and R. Medell´, 'The jaguar's spots are darker than they appear: Assessing the global conservation status of the jaguar Panthera onca', *Oryx*, (2017), 52, 300–315; M. S. Mendoza, P. Cun, E. Horstman, S. Carabajo and J. J. Alava, 'The last coastal jaguars of Ecuador: Ecology, conservation and management implications', in A. B. Shrivastav and A. P. Singh (eds), *Big Cats* (InTech, 2017), pp. 111–131; G. Zapata-Ríos and E. Araguillin, 'Estado de conservación del jaguar y el pecarí de labio blanco en el Ecuador occidental' [State of conservation of the jaguar and the white-lipped peccary in western Ecuador], *Revista Biodiversidad Neotropical*, (2013), 3(1), 21–29.
54 Roy et al, p. 13, relying on A. Castellanos, 'Andean bear home ranges in the Intag region, Ecuador', *Ursus*, (2011), 22(1), 65–73.

55 Roy et al, p. 13, relying on A. Arteaga, R. A. Pyron, N. Penafiel, P. Romero-Barreto, J. Culebras, L. Bustamante and J. M. Guayasamin, 'Comparative phylogeography reveals cryptic diversity and repeated patterns of cladogenesis for amphibians and reptiles in northwestern Ecuador', *Plos One*, (2016), 11(4) doi:10.1371/ journal.pone.0151746; E. E. Tapia, L. A. Coloma, G. Pazmiño-Otamendi and N. Peñafiel, 'Rediscovery of the nearly extinct longnose harlequin frog Atelopus longirostris (Bufonidae) in Junín, Imbabura, Ecuador', *Neotropical Biodiversity*, (2017), 3, 157–167.

56 Roy et al, p. 13, relying on O. Jahn, 'Rediscovery of Black-breasted Puffleg Eriocnemis nigrivestis in the Cordillera de Toisán, north-west Ecuador, and reassessment of its conservation status', *Cotinga*, (2008), 29, 31–39; M. R. Willig and S. J. Presley, 'Biodiversity and metacommunity structure of animals along altitudinal gradients in tropical montane forests', *Journal of Tropical Ecology*, (2016), 32, 421–436. doi:10.1017/s0266467415000589

57 Roy et al, 'New Mining Concessions Could Severely Decrease Biodiversity and Ecosystem Services in Ecuador', op. cit., p. 13, relying on L. Endara, N. H. Williams and S. León-Yañez, 'Patrones de endemismo de orquídeas endémicas Ecuatorianas: Perspectivas y prioridades para la conservación' [Patterns of endemism of Ecuadorian endemic orchids: Perspectives and priorities for conservation], Paper presented at the Proceedings of the Second Scientific Conference on Andean Orchids, Loja, Ecuador, 2009.

58 Wilson, *Biophilia*, p. 85.

59 Roy et al, 'New Mining Concessions Could Severely Decrease Biodiversity and Ecosystem Services in Ecuador', op. cit., p. 9.

60 Ibid.

61 Judgment of 19 June 2019, summing up the *Amicus Curiae* of the biologist Elisa Levy Ortiz, p. 65.

62 Roy et al, 'New Mining Concessions Could Severely Decrease Biodiversity and Ecosystem Services in Ecuador', op. cit., p. 9, relying on B. Ríos-Touma, A. Morabowen, I. Tobes & C. Morochz, 'Altitudinal gradients of aquatic macroinvertebrate diversity in the Choco-Andean region of Ecuador', Oral Presentation, 2017. Paper presented at the Society for Freshwater Science Annual Meeting, Raleigh, NC.

63 Judgment of 19 June 2019, summing up the *Amicus Curiae* of the biologist Elisa Levy Ortiz, p. 65.

64 Ibid.

65 Ibid.

66 The Living Rainforest, 'The vital importance of cloud forests', 3 January 2020.

67 Alexander von Humboldt, *Reise auf dem Río Magdalena, durch die Anden und Mexico*,vol. I: Texte, (Akademie-Verlag, 1986), 358.

68 Andrea Wulf, *The Invention of Nature: Alexander von Humboldt's New World* (Knopf, 2015).

69 Biodiversity Heritage Library, 'Alexander von Humboldt and the Interconnectedness of Nature: Exploring Humboldt's legacy as a Father of Modern environmentalistm', Earth Optimism, 2020, https://blog.biodiversitylibrary.org/2020/10/alexander-von-humboldt.html.

70 Ibid.

71 Ibid.

72 Sarah Sloat, 'In Earth's Vanishing Cloud Forests, "Many Species are going to be lost"', *Inverse*, 17 April 2019, interviewing Eileen Helmer PhD, a research ecologist with the US Forest Service and lead author of a paper on climate change as a major threat to cloud forests..

73 Dacher Keltner, *Awe: The Transformative Power of Everyday Wonder* (Allen Lane, 2023), p. xvi.

74 Ibid, p. xxv.

75 Ibid.

76 Ibid, p. xvi.

77 Ibid, p. 127.

78 The Magic of Rights: Critical Reflections on Rights of Nature, Environmental Rights and Related Concepts, 25-26.01.2024, Max Planck for Social Anthropology.

79 Monica Feria-Tinta, *Amicus Curiae*, Los Cedros case, http://esacc.corteconstitucional.gob.ec/storage/api/v1/10_DWL_FL/e2NhcnBldGE6J2VzY3JpdG8nLCB1dWlkOidkNTNjM2VkOC-03ODIyLTQwODYtYTM5YS1lNzU4NjU1OWFhOTEuc-GRmJ30=.

80 Wilson, *Biophilia*, op. cit., p. 19.

81 As noted by Pierre-Marie Dupuy, Ginebra Le Moli and Jorge E. Viñuales, *Customary International Law and the Environment*, C-EENRG Working Papers 2018-2, December 2018, p. 9.
82 United Nations General Assembly Res. 37/7, 28 October 1982.
83 Ibid.
84 Philippe Sands and Jacqueline Peel, *Principles of International Environmental Law* (Cambridge University Press, fourth edition, 2018), p. 37.
85 Wilson, *Biophilia*, op. cit., p. 13.
86 Christophe D. Stone, *Should Trees Have Standing?* (OUP, 2010) (third edition).
87 Ibid. See, for example, reference to Nature as 'natural resources', (p. 117); as 'one form of capital stocks', a 'stock of resources' (p. 110); 'assets' and 'wealth' (pp. 110, 121). See also the approach to Deep Sea Mining: 'There are also minerals, which will presumably become more accessible as extraction technology improves', p. 96.
88 See, for example, reference to Nature as a stock of planetary 'goods and services' for future generations, p. 111, and 'copying and exploiting genetic information' in his considerations on bio-diversity, p. 136.
89 See ibid, p. 87.
90 Constitutional Court of Ecuador, Case No 1149-19-JP/20, Judgment No. 1149-19-JP/21, 10 November 2021 (Los Cedros judgment).
91 Ibid, para. 35 [Author's translation].
92 Ibid, para. 34 [Author's translation].
93 Ibid, para. 48 [Author's translation].
94 Ibid, para. 43 [Author's translation].
95 Ibid, para. 48 [Author's translation].
96 Gabriel García Márquez, *One Hundred Years of Solitude* (Avon Books, 1971), p. 219.
97 Işıl Şahin Gülter, 'The Capitalocene and Slow Violence in Gabriel García Marquez' One Hundred Years of Solitude', *Journal of Narrative and Language Studies*, Special Issue (2023), 11(21), p. 17.
98 Gabriel García Marquez, *El General en su Laberinto* (Editorial Oveja Negra, 1989), p. 97. [Author's translation].
99 Los Cedros Judgment, para. 52 [Author's translation].

100 Ibid, para. 68 [Author's translation].
101 Ibid, para. 68 [Author's translation].
102 Ibid, para. 69 [Author's translation].
103 Spinoza, *Ethics* (J. M. Dent & Sons, 1941), see Preface, Part III, 'On the Origin and Nature of the Affects', p. 84.
104 Ibid, para. 344(c) [Author's translation].

10. Into the Deep Blue

1 LuAnn Dahlman and Rebecca Lindsey, 'Climate Change: Ocean Heat Content', 1 August 2018, https://www.climate.gov/news-features/understanding-climate/climate-change-ocean-heat-content.
2 Article 138 of the Rules of the International Tribunal for the Law of the Sea.
3 See IPCC Sixth Assessment Report, Climate Change 2022: Impacts, Adaptation and Vulnerability (Chapter 3: Oceans and Coastal Ecosystems and their Services), Executive Summary, https://www.ipcc.ch/report/ar6/wg2/chapter/chapter-3/.
4 See Nilufer Oral, 'Climate Change and protecting the oceans: A Tale of two regimes', 11 May 2018; IUCN D. Laffoley and J. M. Baxter, 'Explaining Ocean Warming: Causes, scale, effect and consequences', citing the Fifth Assessment Report published by the Intergovernmental Panel on Climate Change, p. 17.
5 See https://www.un.org/en/climatechange/science/climate-issues/ocean-impacts.
6 WWF, 'Climate, Nature and Our 1.5°C Future Report, A Synthesis of IPCC and IPBES Reports', pp. 12–13.
7 IPCC. Climate Change 2023, Sixth Assessment Report, Synthesis Report, Summary for Policymakers, a Report of the Intergovernmental Panel on Climate Change, (Hereafter 'IPCC, AR6 SYR'), p. 5.
8 United States Government Official Site, National Ocean's Service, 'What is the cryosphere?', https://oceanservice.noaa.gov/facts/cryosphere.html#:~:text=The%20cryosphere%20is%20the%20frozen%20water%20part%20of%20the%20Earth%20system.&text=This%20includes%20frozen%20parts%20of,temperatures%20below%200°C.

9 WWF, 'Climate, Nature and Our 1.5°C Future Report', p. 12.
10 IPCC. AR6 SYR, p. 5.
11 Ibid.
12 Ibid.
13 Tanaka Yoshifumi, 'Article 1', in Alexander Proelss (ed.), *United Nations Convention on the Law of the Sea, A Commentary* (C. H. Beck, Hart Nomos, 2017), p. 23.
14 Edward O. Wilson, *Half-Earth* (Liveright, 2016), book jacket.
15 M. M. Meredith, S. Sommerkorn, C. Cassotta, A. Derksen, A. Ekaykin, G. Hollowed, A. Kofinas, J. Mackintosh, M. M. C. Melbourne-Thomas, G. Muelbert, H. Ottersen, H. Pritchard and E. A. G. Schuur, 'Polar Regions', in: *IPCC Special Report on the Ocean and Cryosphere in a Changing Climate*, H.-O. Pörtner, D.C. Roberts, V. Masson-Delmotte, P. Zhai, M. Tignor, E. Poloczanska, K. Mintenbeck, A. Alegría, M. Nicolai, A. Okem, J. Petzold, B. Rama and N. M. Weyer (eds.), (Cambridge University Press, 2019), pp. 203–320, p. 205, https://doi.org/10.1017/9781009157964.005.
16 Ibid, p. 206.
17 See IPCC Sixth Assessment Report, Chapter 9: Ocean, Cryosphere and Sea Level Change.
18 WWF, 'Polar Bear: A Powerful Predator on Ice Species', https://www.wwf.org.uk/learn/wildlife/polar-bears#:~:text=Previous%20Next-,Climate%20change,ice%20to%20raise%20their%20young.
19 Mary Jane Schramm, 'Tiny Krill: Giants in Marine Food Chain', *National Marine Sanctuary Program*, https://sanctuaries.noaa.gov/news/features/1007_krill.html#:~:text=Krill%20is%20the%20near%2Dexclusive,the%20entire%20marine%20ecosystem%20suffers.
20 Ibid.
21 Emily S. Klein et al, 'Impacts of rising sea temperature on krill increase risks for predators in the Scotia Sea', *Plos One* Journal Open Access, 31 January 2018, https://doi.org/10.1371/journal.pone.0191011.
22 S. Kawaguchi et al, 'Risks Maps for Antarctic Krill under Projected Southern Ocean Acidification', *Nature Climate Change*, 7 July 2013, https://www.nature.com/articles/nclimate1937. Paradoxically, the contribution of the Antarctic krill to blue carbon processes is

significant. Scientists have found that when krill excrete faeces, this carbon is locked away in the deep sea, helping to maintain stable climatic conditions that are beneficial for humanity. A report found that these processes sink the equivalent of 23 megatonnes of carbon annually in just one area of the Southern Ocean. See WWF, 'Antarctic Krill: Powerhouse of the Southern Ocean', Report, 2022; Haley Dunning, 'Tiny Antarctic creatures provide US$8.6 billion of carbon storage via their poo', Imperial College News, 20 October 2022.

23 Oliver Milman, 'World Faces "terminal" loss of Arctic sea ice during summers, report warns', *Guardian*, 7 November 2022.

24 Ibid. *See* International Cryosphere Climate initiative, "The State of the Cryosphere 2022, Growing Loses, Global Impacts", November 2022 p. 2.

25 Craig Kielburger, 'Inuit activist Sheila Watt-Cloutier connects the dots on climate change', https://www.we.org/en-GB/we-stories/opinion/craig-column-inuit-activist-sheila-watt-cloutier-climate-change#:~:text=Think%20about%20the%20interconnectedness%20of,world%2C%20and%20it%27s%20breaking%20down.

26 Article 194 (1).

27 For a definition of 'the Area', see N. Oral, 'The Common Heritage of Mankind under International Law', in T. Campanella (ed.), *Handbook of Seabed Mining & The Law of the Sea* (Routledge, 2023), pp. 33–37.

28 J. Dingwall, 'Commercial Mining Activities in the Deep Seabed Beyond National Jurisdiction: The International Legal Framework', in C. Banet (ed.), *The Law of the Seabed* (Brill, 2020), p. 137.

29 US Government Accountability Office (GAO), 'Science and Tech Spotlight: Deep-Sea Mining', 15 December 2021, available online https://www.gao.gov/products/gao-22-105507 [Accessed 12 April 2022].

30 As of 12 April 2022, the following contractors had been awarded ISA Exploration Contracts: Interoceanmetal Joint Organization, JSC Yuzhmorgeologiya, Government of the Republic of Korea, China Ocean Mineral Resources Research and Development Association, Deep Ocean Resources Development Co. Ltd, Institut français de recherche pour l'exploitation de la mer, Government

of India, Federal Institute for Geoscience and Natural Resources, Nauru Ocean Resources Inc., Tonga Offshore Mining Ltd, Global Sea Mineral Resources NV, UK Seabed Resources Ltd, Marawa Research and Exploration Ltd, Ocean Mineral Songapore Ptd. Ltd, Cook Islands Investment Corporation, China Minmetals Corporation, Beijing Pioneer Hi-Tech Development Corporation, Ministry of Natural Resources and Environment of the Russian Federation, Government of Poland, Japan Oil, Gas and Metals National Corporation, Companhia de Pesquisa de Recursos Minerais S.A., and Blue Minerals Jamaica Ltd.

31 H. Reid, 'Pacific Island of Nauru sets two year deadline for U.N. deep-sea mining rules', *Reuters*, 29 June 2021, available online https://www.reuters.com/business/environment/pacific-island-nauru-sets-two-year-deadline-deep-sea-mining-rules-2021-06-29/ [Accessed 29 September 2021].

32 K. M. Gjerde, 'Current challenges regarding deep-sea mining and protection of ocean life beyond national boundaries', Lauterpacht Centre for International Law, Public Lecture, 11 February 2022.

33 United State Geological Survey (USGS), Official Website of the US Government, https://www.usgs.gov/faqs/what-are-gas-hydrates [accessed 5 September 2021].

34 USGS, Official Website of the US Government, https://www.usgs.gov/faqs/what-are-gas-hydrates [accessed 5 September 2021].

35 Methane hydrate, https://www.sciencedirect.com/topics/earth-and-planetary-sciences/methane-hydrate [accessed 2 January 2022].

36 H. M. King, 'Methane Hydrate', *Geoscience News and Information*, available online https://geology.com/articles/methane-hydrates/ [accessed 2 January 2022].

37 World Ocean Review, 'Methane Hydrate', https://worldoceanreview.com/en/wor-3/methane-hydrate/mining-impacts/, [accessed 2 January 2022].

38 Ibid.

39 D. Yan et al, 'Governing the transboundary risks of offshore methane hydrate exploration in the seabed and ocean floor – an analysis on international provisions on Chinese law', *Journal of World Energy Law and Business*, (2020), 13(2), 185–203, p. 186.

40 World Ocean Review, 'Methane Hydrate', op. cit.
41 Article 192.
42 See Commentary on Article 192 in A. Proelss (ed), *United Nations Convention on the Law of the Sea: A Commentary*, op. cit., p. 1280.
43 WWF submissions before ITLOS, Case No 31, para. 113.
44 Ibid, para. 114. In particular, under Article 192 and Article 145 of the Convention on the Law of the Sea.
45 Wilson, *Half-Earth*, op. cit., p. 129.
46 Ibid.
47 Ibid, p. 130.
48 Melina Buckley, Review, 'The Pearl Button', 2015, https://agoodmovietowatch.com/the-pearl-button-2015/.
49 The Voyager Golden Records are two identical phonograph records that were included aboard the two Voyager spacecraft launched in 1977. One of the purposes was to send a message to extra-terrestrials who might find the spacecraft as it journeyed through interstellar space.
50 Jaqueline Garget, 'Disappearing notes in classical tune highlight dramatic loss of Humpback Whales', 14 October 2022 (blog, University of Cambridge Website) https://www.cam.ac.uk/stories/Hebrides-Redacted.
51 Ibid.
52 'Patricio Guzman's *The Pearl Button:* 'We are all streams from one water', Gerry in Film, 13 April 2016, https://gerryco23.wordpress.com/2016/04/13/patricio-guzmans-the-pearl-button-we-are-all-streams-from-one-water/.
53 Ibid.
54 NASA, 'Comet provides new clues to origins of Earth's Oceans', 23 May 2019.

Epilogue

1 Professor Slavoj Žižek, Full Address and Q&A, Oxford Union, https://www.youtube.com/watch?v=545x4EldHlg.
2 Eloise Barry, 'Why Chileans Rejected a New, Progressive Constitution', *Time*, 5 September 2022.
3 Edward O. Wilson, *Biophilia* (Harvard University Press, 1984), p. 25.

4 Wilson, *Biophilia*, p. 139.
5 Ibid, p. 121.
6 Edward O. Wilson, *Half-Earth* (Liveright, 2016).
7 Julian of Norwich, *Revelations of Divine Love*, OUP, 2015, p. 20.
8 Like Julian of Norwich, the mystic, learned to see in her revelations.
9 Schedule 5, Section 9.

Acknowledgements

This book owes a lot to a number of people. I am grateful to Klemens Felder, my partner, without whom I probably would have never learnt to pay attention. He inspired me with his love for Nature and nurtured my wonder. Having been a keen climber in the Alps and studied forestry in Vienna, he has been a patient teacher. He has taught me his love for mountains, trees, forests, birds and all kinds of wildlife through decades of being in direct contact with it all.

I am grateful to Barry Cheetham who first read what I wrote and encouraged me.

I owe a great deal to Sophie Scard, my agent, without whom this book would have never seen the light.

I am grateful to Fiona Crosby, my editor at Faber, who has been instrumental in bringing this book to life. From the outset, the Faber team treated this book as a precious item for which I am so thankful. I am indebted to the many individuals at Faber who contributed to the publication process and who paid so much attention to detail in this process.

I am indebted to my family and to my roots which helped me to understand perspectives of Nature and the world, which this book honours.

This book is in memory of Antonio 'loco' Sánchez, my literature teacher at secondary school, who always knew I'd be a writer.